The 360 Manual of Knowledge

By MichaEl Bey(Mic Henchmen)

Bringing the doctrine of "Teachnology" and "Lifeology"

The Universe Exist Within You.

Wombman,Man, and Child
(120) Knowledge,(120)Wisdom, and(120)Understanding
(360 Degrees)

Understand The "Golden Means"

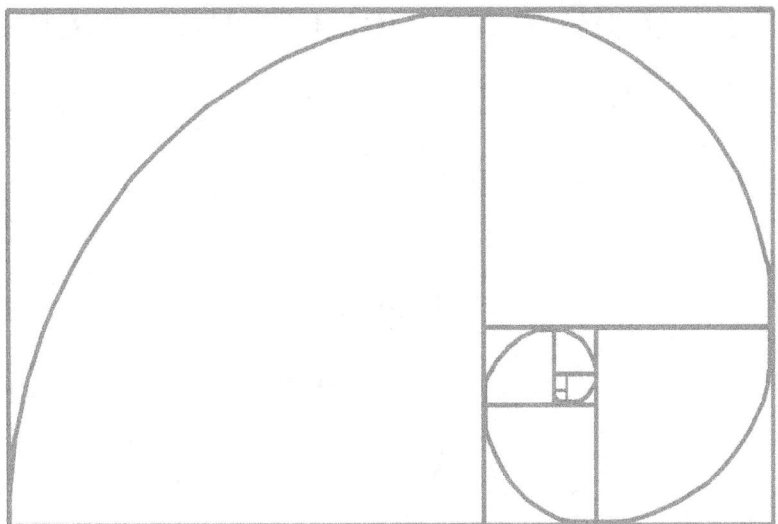

INTRODUCTION

The whole purpose of this manual is to guide you to understanding of self. It will break down basic principles on different subject matters. It is to give the people a basic foundation on what to research, read, and study. This will put you on the path to determine "right knowledge" from "wrong knowledge". You must look outside the box square from all four angles, because everyday life is a lesson learned from what you experience, so get in tune with your inner self,become conscious and over stand this basic guide.

**Read
Research
Study
&
"The Truth Shall Set You Free"**

How to used the 360 Manual of Knowledge

Research,Study,Cross Reference and Study what you Research.
Just don't take may words and information or knowledge, find out for self in the way it make sense to you. Each person has a journey in life, and each person must understand the knowledge that they come into, for the way it can help benefit them on their journey in life in the physical form, But the idea is to
become more conscious in this physical realm. Never let no one dictate how you see the future of your journey, because you is the captain of your own ship as being a true navigator guided by spiritual,cosmic energy of conscious in the universe that is within you.

Here is a list of scholars to research their study of work.

These Scholars are Grand Master Teachers who has paid the way for the people of today.

1.Dr. Yosef Ben Jochannan- He was a legendary Great Grand Master Historian of history. His Information is dealing Ancient Kemet, Sumerians,Moors,The Americas,Africa and Europe. His work speak for it self, and we today use his scholarship works in this day of time.
2.Noble Drew Ali-His work was to uplift humanity of bringing the culture of Moorish Science to the people, so they could understand their truth self on their land they descend from. He brought fourth nationality principles, so the Indigenous & Native people of America, had a foundation of knowledge to reclaim their sovereignty.
3.Dr. John Henry Clark-His scholarship work of history has serve the community and the world with great influence, his work of history of America,Europe and Africa is inspirational information that the people need to absorb.
4.Dr. Ivan Sertima- He was big time Historian on Moorish history,he has many books about the Moors in how they played the role in this society today. One of his great writings is the "The Golden of the Moors"and many others.
5.J.A. Rogers-His scholarship of work was dealing with Historical facts ,Racism and how it became. Also he has a wide range of American History in pertaining to the first so called Black president in America. His work was express threw the early 1900s, and is still being recognize today in the 21st century.
6.Dr. Sebi- He was a Herbalist,Biochemist, and Naturalist, his legendary works of Health was teaching the world how to eat on a Alkaline diet,and that it is mucus which cause the disease. He has heal over many people from almost every disease. And his work is still to be continue.
7.Mantak Chia- He is a Grandmaster of the chi energy, which deals with breathing techniques to help aid in healing the body. An expert on the human anatomy, to help connect people to a higher stage of consciousness. His work has been express all over the world in today society.

1. When it comes to history, law, and language, it is very necessary to do the etymology on certain words you may come across, so you will know its origin and it will put things in its proper perspective.

2. When comes to Gematria, it is best to used that in decoding the riddles what these secret societies and corrupt government put out. It can be used in many ways such as connecting the dots of Astrology and revealing how the universe operates. Understand it is certain numbers that is used for their communication in how they continue to control the masses.

3. When it comes to health, we need to understand the herbs that heals the body. We need to understand breathing techniques, and its benefits. The practice of activated kundalini energy, is very key to opening your mind and consciousness.

4. Kemetic, Moorish Science and Masonry are the based foundation of culture science that is been used today. To having a better understanding of Masonry, you need understand Moorish Science, and to have a better understanding of Moorish Science.

You will need to have better understanding of Kemetic Science because majority of the world knowledge comes out of Ancient Kemet, and we as a people will be a fool not to recognize that as a historical fact.

Research, Study, Cross Reference and Study what you Research.

The 360 Manual of Knowledge

Chapter 1
Increase Your Health

The society we live in today, market over 80% of foods that is unnatural and has chemicals to enhance the taste and preserve it for more shelf life to be sold to the people. So, If that is the case, we as people just can't settle on a diet fill with toxins
and chemicals. The diet we eat is one of the main reason people in the world sulfur
from sickness and diseases, which can be prevented and heal ourselves with natural herbs and foods that alkalizes the body.

Now we need to understand **"Alkalinity and Acidity"**

The body has "**Ph balance scale**"from **"0 to 14"** seven is neutral, anything under **"7"** is **(Acidity)** meaning the body is doesn't have enough of a **electrical spark** and the cells lack oxygen and is vulnerable to an infection or disease. Anything over **"7"** is **(Alkalize),** means the body is now more electrical, and it has the proper amount of oxygen and can neutralize any harmful germ or disease.

Note: No germ such as a virus,fungus,bacteria, protist, or parasite, cannot survive in an alkalize environment. So since we understand that concept, that is the base formula to healing any disease. You can boast your ph a balance with baking soda(Sodium Bicarbonate) be sure to get ph balance test kit and keep check on your ph balance

The body ph balance Alkalize level is to be between (7.0 & 9.6)

This is the basic principle of healing the body.

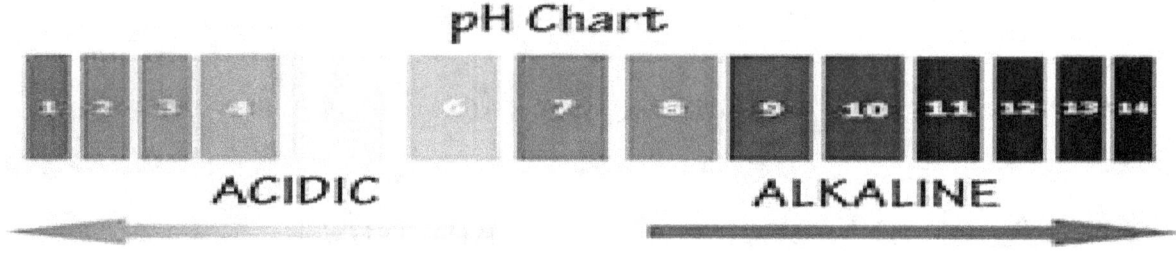

pH Chart

ACIDIC   ALKALINE

Now According to **"Dr. Sebi"**years of research and studies,
"Mucus" in the cell membrane is the number one cause of a disease.
So on that note, by getting rid of the mucus, which makes the body acidity and keeping the body alkaline, you have the power to reverse any disease, by following a protocol diet.

Have you ever thought about why western doctors are quick to prescribe you a drug and not herbs to treat or heal a disease. We must understand that they make billions of dollars every year off these prescription drugs, but you still remain sick with the disease. When we look at the situation, some things just don't add up. It is a agenda deeper than money, it is an global depopulation agenda.
The ideal is not to heal you, but treat the symptoms, so you can keep coming back to buy the drugs and slowly destroy your body with these chemicals or drugs during the process. So we as the public people must take control of our body, and be willing to not depend on medical doctors and prescription drugs for our health problems. My idea based upon the diet in society today, is to balance your diet, and do

some research on what you are eating. You are what you eat, for example, if you continue to eat fast food everyday which is full with toxins, your body will be full of toxins and become very acidity. We as people must have some type of conscious with our diet, it is very critical that we do so.

Nutritional Guide of Dr. Sebi The healing diet is found below. It's important to keep in mind that "Dr. Sebi has recommended the foods that are listed here for the reversal of disease for over 30 years. If your favorite food is missing from the list, Research and results have proven that it has no nutritional value and may be detrimental to your health.

Here is a list of alkaline Foods according to the Dr. Sebi list

Vegetables •
Amaranth greens – same as Callaloo, • Avocado • Bell Peppers • Chayote (Mexican Squash) • Cucumber • Dandelion greens • Garbanzo beans • Green banana •I zote – cactus flower/ cactus leaf – grows naturally in California • Kale • Lettuce (all, except Iceberg) • Mushrooms (all, except Shiitake) • Nopales – Mexican Cactus • Okra • Olives • Onions • Poke salad – greens • Purslane (Verdolaga) • Sea Vegetables (wakame/dulse/arame/hijiki/nori) • Squash • Tomato – cherry and plum only • Tomatillo • Turnip greens • Watercress • Zucchini

Fruits
(No canned or seedless fruits) • Apples • Bananas – the smallest one or the Burro/mid-size (original banana) • Berries – all varieties- Elderberries in any form – no cranberries • Cantaloupe • Cherries • Currants • Dates • Figs • Grapes- seeded • Limes (key limes preferred with seeds) • Mango

• Melons- seeded • Orange (Seville or sour preferred, difficult to find) • Papayas • Peaches • Pear • Plums • Prickly Pear (Cactus Fruit) • Prunes • Raisins –seeded • Soft Jelly Coconuts • Soursops – (Latin or West Indian markets) • Tamarind

Herbal Teas
• Allspice • Anise • Burdock • Chamomile • Elderberry • Fennel • Ginger • Raspberry • Tila Spices and Seasonings Mild flavors • Basil • Bay leaf • Cloves • Dill • Oregano • Parsley • Savory • Sweet Basil

• Achiote • Cayenne/ African Bird Pepper • Coriander (Cilantro) • Habanero • Onion Powder • Sage Salty Flavors • Pure Sea Salt • Powdered Granulated Seaweed (Kelp/Dulce/Nori – has "sea taste") Himalaya Pink Salt , it is the purest salt

Sweet Flavors • 100% Pure Agave Syrup – (from cactus) • Stevia Powder

These are some herbs that I suggest.
1.Irish Sea Moss- contains over 92 minerals, and treats and heal many diseases
2.Black Seed Oil-help reduce high fevers, and treats and heal many diseases
3.Sarsaparilla-Has high content of iron, which attracts other minerals in the body
4.Ginkgo Biloba- help increase energy to the brain, increase brain capacity, & enhance memory
5.Fenugeek- helps in reducing inflammation, and controlling blood sugar levels
6.Chaparral- builds up the immune system and helps in reducing inflammation
7.Kelp- Has high content of iodine, which helps balance the thyroid gland

Foods that act as toxins

1.Anything that has gluten
2.Artificial sweeteners such as High fructose Corn Syrup, Aspartame, Sucrose and Table Sugar
3.Fast Foods and Junk Foods
4.Sodas and Soft drinks such as Pepsi,Coca Cola and Dr. Pepper. t ac "They contain the Artificial sweeteners.
5.White Flour
6.Hydrogenated & Partial Hydrogenated oils
7.Process foods
8.Pork
9.Corn

Foods that act like foods

1.Spinach
2.Carrots
3.(Beans) but not stream beans they are highly acidity
4.Sweet potatoes
5.Beets
6.Chicken, turkey, sea foods if not process.
7.Almonds
8.Cashews
9.Macadamias

Foods and herbs that act like medicine

1.All Natural Herbs & Spices
2.All Natural Fruits and Vegetables
3.All Natural Nuts
4.All Natural Oils

The ideal is to create balance with your diet, eat less foods that act as toxins, and strive to eat foods that act as foods and medicine.

1.Detox the body
2.Alkaline the body
3.Feed the body minerals and not toxins
4.Exercise at your own speed

The information contained in these topics is not intended nor implied to be a substitute for professional medical advice, it is provided for educational purposes only. You assume full responsibility for how you choose to use this information. Always seek the advice of your physician or other qualified healthcare provider before used.

Note:Do not used any prescription drugs, while taking these herbs, consult your local doctor or physicians for more advice. You only used the herbs to supplement and to help heal the body.

Key things to know, is Iron is the only mineral in the body which is magnetic, so when the cells in the body gets iron, the cells in the body flow thru the bloodstream and attracts other minerals increasing your Health. Iron is a necessary mineral needed in the body and many people are Iron deficiency, which makes them a anemic, and the body don't have enough red blood cells to carry oxygen thru out the body.

Herbs high in Iron -Sarsaparilla-Irish Sea Moss

Remedies and Protocol how to heal Diseases

(1.Alkaline The Body)

Cancer- 1tbsp of Baking Soda(Sodium Bicarbonate) In a cup of Spring or Distilled Water, 3 times a day, ***Be sure to get a **ph balance test kit,** you don"t want to raise ph balance past **9.6** it will cause the body to have severe problems. **(7.0 to 9.6) ph balance**

1 tbsp of **Black Seed oil** mixed with a half cup of lime juice, in a cup of Spring or Distilled water

1 tbsp of **Irish Sea Moss** In a cup of Spring or Distilled Water, 2 times a day.

1 tbsp of **Gogi Berry Juice or Powder,** In a cup of Spring or Distilled Water, 3 times a day.

5 to 7 **Apricot seeds** 2 to 3 times a day. **Note:It contain Vitamin B17, which directly kills cancer cells, be sure to begin with small dosages**

All foods that act as Toxins must be remove from the diet.

Alzheimer's -1 tbsp of **Ginkgo Bilbao** In a cup of Spring or Distilled Water, 1to 2 times a day.
1 tbsp of **Gotu Kola** In a cup of Spring or Distilled Water, 1 to 2 times a day.

Arthritis – 1 tbsp **Fenugeek or Chaparral** helps reduce inflammation to help relieve the pain, In a cup of Spring or Distilled Water, 1 time a day and
1 tbsp of I**rish Sea Moss** In a cup of Spring or Distilled Water, 1 time a day. Eat foods high in Calcium and Phosphorus. **Note:Get plenty of sunlight, the body transform the energy from the sun to vitamin d, which helps assist calcium to help grow and rebuild bones.**

Lupus and Aids- The body is so acidity, that the cells in the body start to become confuse. The body now lacks oxygen and the cells start to act abnormal.

1 tbsp of **Baking Soda (Sodium Bicarbonate)** In a cup of Spring or Distilled Water, 2 to 3 times a day, 1 tbsp of Irish Sea Moss In a cup of Spring or Distilled Water, 1 time a day
The key is to alkaline the body and remove the toxins.(Raise ph balance) **(7.0 to 9.6)**

Diabetes Type 1 and 2-Note:Its all about the diet. No foods that act as toxins, No meats,
and starchy foods. Raw fruit,vegetable and Nut diet. A lot of green leaf vegetables. No high sugar fruits such as oranges,watermelon,pineapples and grapes. Etc. If **"Type 1"** if sugar level may drop, I suggest you bring sugar level up with the natural sugars in the sweet fruits. In order to heal diabetes, you must have a very strict diet on more alkaline foods. The process of healing can take from "**21 days to 3 months**", if you if following the protocol diet and not cheating yourself.

Herb supplement:You can choose which herb supplement you want.
A. 1 tbsp of Black Seed oil and lime juice In a cup of Spring or Distilled Water, 2 times a day. And 1 tbsp of Goji Berry juice In a cup of Spring or Distilled Water, 1 time a day
B.1 tbsp of Chaparral In a cup of Spring or Distilled Water, 2 times a day. And 1 tbsp of Goji Berry juice In a cup of Spring or Distilled Water, 1 time a day
C.1 tbsp of Chaparral In a cup of Spring or Distilled Water, 2 times a day. And 1 tbsp of Goji Berry juice In a cup of Spring or Distilled Water, 1 time a day
D. 1 tbsp of Irish Sea Moss, In a cup of Spring or Distilled Water, 1 time a day And 1 tbsp of Fenugeek In a cup of Spring or Distilled Water, 1 time a day
E.1 tbsp of Chaparral In a cup of Spring or Distilled Water, 2 times a day. Eat 5 to 8 raw olives a day.

Diabetes is all in the mind, you as a person must want to let got the poor diet, and return to a more natural diet to heal yourself, It all depends on you as a person if you ready for that challenge, if not you will continue to sulfur from the disease and have problems in the future. You is the soul controller, **Note:You Control your Diet and Don't let your diet control you.**

Asthma-
1 tbsp of Chaparral In a cup of Spring or Distilled Water, 2 times a day.
1 tbsp of Fenugeek In a cup of Spring or Distilled Water, 2 times a day
1tbsp of Ortiga(Stinging Needle) In a cup of Spring or Distilled Water, 2 times a day
1 tbsp of Bio Ferro (Dr.Sebi Product) 1 time a day

Remove all foods that acts as toxins

Alkaline the body ph balance to **(7.0 to 9.6)**

Obesity-Kelp is a type of seaweed that's rich in antioxidant vitamins and iodine. It stimulate a
hormone produced by the thyroid gland that's responsible for boosting metabolism, so you'll burn more calories by the hour. Kelp is very useful for thyroid-related obesity.

1 tbsp of kelp 2 times a day In a cup of Spring or Distilled Water, 2 times a day.

1 tbsp of Irish Sea Moss 2 times a day In a cup of Spring or Distilled Water, 2 times a day.

Remove all junk food and starch from your diet."Foods that acts as toxins"

Drink plenty of water (Spring or Distilled) Aerobic exercise at your own speed

High Blood Pressure-Remove all table salt from diet. Add Himalaya Pink Salt or Sea Salt.
(Table salt) is inorganic and contains 1/3rd Sodium,1/3rd glass/ & 1/3rd sand. The glass and sand in the table salt, tears threw the arteries and makes it bleeds,causing cholesterol to seal up the wounds, which cause build up in the vessels. That is when the blood pressure start become unbalance and pump abnormal.
Add more green leaf vegetables to your diet. Follow the protocol for 3 months for full effect of healing

Note:Do used any blood pressure or blood thinning prescription drugs, while taking these herbs, consult your local doctor or physicians for more advice.

1 tbsp of Apple Cider Vinegar with "the mother" with spring or distilled water 3 times a day
1 tbsp of Gogi Berry Juice 3 times a day
1 tbsp of Black Seed Oil with spring or distilled water 2 times a day
1 tbsp of Irish Sea Moss with spring or distilled water 2 times day

Crohn's Disease – It is a digestive problem like diabetes, the idea is to reduce the inflammation and remove all foods that acts as toxins.
1 tbsp of Chaparral In a cup of Spring or Distilled Water, 2 times a day. And 1 tbsp of Goji Berry juice In a cup of Spring or Distilled Water, 1 time a day
1 tbsp of Chaparral In a cup of Spring or Distilled Water, 2 times a day. And 1 tbsp of Goji Berry juice In a cup of Spring or Distilled Water, 1 time a day
1 tbsp of Irish Sea Moss, In a cup of Spring or Distilled Water, 1 time a day And 1 tbsp of Fenugeek In a cup of Spring or Distilled Water, 1 time a day
1 tbsp of Chaparral In a cup of Spring or Distilled Water, 2 times a day. Eat 5 to 8 raw olives a day.

Chapter 2
Gematria- The Math Code system of words

The form of **Gematria** can be traced back to the **Kabbalah**. It is a **language** of numbers coded in **words** and by using this system of math can help you decipher **truth behind the words**. So if you understand basic Arithmetic, you will be able to advance on a good note.

Simple English Gematria

Simple English Gematria Chart:

A=1	J=10	S=19
B=2	K=11	T=20
C=3	L=12	U=21
D=4	M=13	V=22
E=5	N=14	W=23
F=6	O=15	X=24
G=7	P=16	Y=25
H=8	Q=17	Z=26
I=9	R=18	

Each Letter is in alphabetical order from (A to Z) from (1 to 26) so (A) will be 1, (B) will 2,(C) will be 3,...etc. The numbers can only be reduce to a single digit number. Note: Always isolate the Zero (0)

Now for example: (**A m e r l c a**) Simple Gematria Equals:v**50**
    1 13 5 18 9 3 1   *so you have 1+13+5+18+9+3+1=**50***
                        Now reduce (50) 5+0=(5)

U S A *Simple English* **system equals 41 (21+19+1), which reduces to 5**
21 19 1   **Reduce (4+1)=5**

English Gematria

A	B	C	D	E	F	G	H	I	J
6	12	18	24	30	36	42	48	54	60

K	L	M	N	O	P	Q	R	S
66	72	78	84	90	96	102	108	114

T	U	V	W	X	Y	Z
120	126	132	138	144	150	156

Now in English Gematria Every letter is Added by **6** such as (A) will be **6**, (B) will be **12**,(C) will be **18**,...etc.

(LOVE)in thevEnglish Gematriav system equals 495v(30+60+400+5), which reduces to 18, which reduces to 9

L O V E
30 60 400 5   Reduce (3+6+4+5)=18, now take (18) 1+8=**(9)** which is

the number of completion

Pythagorean English Gematria

In this form of gematria, is more advance,because certain letters has a value of two numbers, From (A to I) will be (1 to 9) Now from (J to R) is (1 to 9) accept (K) it will be (2 or 11) and (S to Z) will be (1 to 9) accept (S will be 1, or 10) and (V will be(4 or 22)

Pythagorean English Gematria Chart:

A=1	J=1	**S=1/10**
B=2	**K=2/11**	T=2
C=3	*L=3*	*U=3*
D=4	M=4	**V=4/22**
E=5	N=5	W=5
F=6	*O=6*	*X=6*
G=7	P=7	Y=7
H=8	Q=8	Z=8
I=9	*R=9*	

1	2	3	4	5	6	7	8	9
A	B	C	D	E	F	G	H	I
J	K	L	M	N	O	P	Q	R
S	T	U	V	W	X	Y	Z	

Now for Example: the word **"mental"** in the *English Gematria* system equals 326 (40+5+50+200+1+30), which reduces to 11, which reduces to 2

11 5 4 5 5 2

(11+5+4+5+2=(22) which is a master number, and does not suppose to be reduce, such as any number as (11,22,33,44,55,66,77,88,99) but it is a acceptation to that rule to reduce if necessary.

Phalanges Gematria

It is a system of gematria that connects directly to (**Astrology** and **Astronomy**) of breaking down the the **14** phalanges of each hand
which sums up to **28**.The evidence is clear that the human body always was connected with the stars,planets and the universe. So if you really want to advance in this line of study in gematria, this is the best of the best.

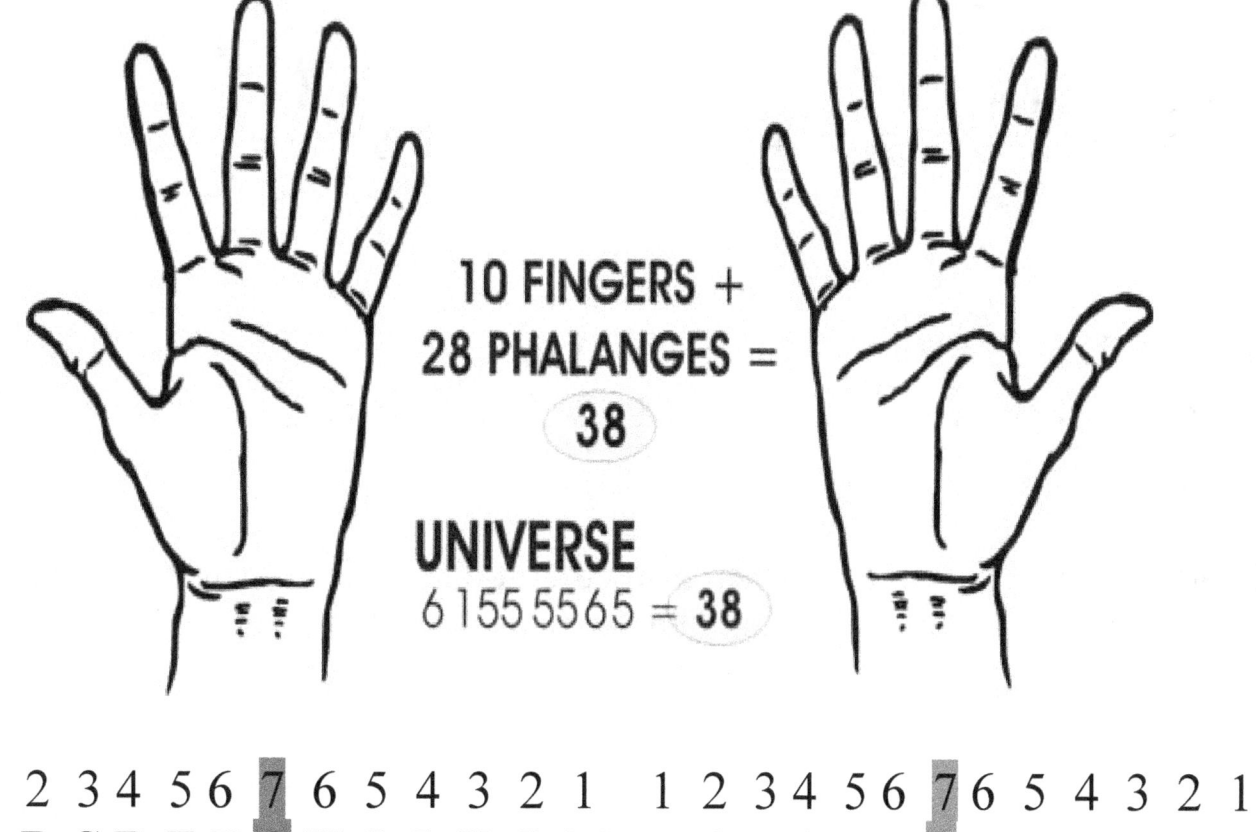

10 FINGERS +
28 PHALANGES =
38

UNIVERSE
6 15 5 5 5 6 5 = 38

1 2 3 4 5 6 **7** 6 5 4 3 2 1 1 2 3 4 5 6 **7** 6 5 4 3 2 1
A B C D E F **G** H I J K L M N O P Q R S **T** U V W X Y Z

Now in this form of gematria it go (1 to 7) such as 1,2,3,4,5,6,(**A to F**) (then the letter (**G**) is (**7**) then counting back, 6,5,4,3,2,1 (**H to M**) that is the first set of the (**14 phalanges** on the **hand**) Now the second set go (1 to 7) such as 1,2,3,4,5,6 (**N to S**) (**T**) is (**7**) then counting back, 6,5,4,3,2,1

Now lets take (**28**) phalanges on the hand(**Twenty Eight, Mother and Father**) used this for example and see were we get.

Twenty Eight
745172 55767 (7+4+5+1+7+2)=**26**   (5+5+7+6+7)=**30**   (26 +30)=**56**
(26)   (30)

Now lets Note:Its **56 constellations** in the south hemisphere, in the Milky Way galaxy,This is a small sample, in how it connects, **Astrology** and **Astronomy.**

Mother Father
127655   617655 (1+2+7+6+5+5)=**26**   (6+1+7+6+5+5)=**30**
(26)   (30)

1 2 3 4 5 6 7 6 5 4 3 2 1   1 2 3 4 5 6 7 6 5 4 3 2 1
A B C D E F G H I J K L M   N O P Q R S T U V W X Y Z

Note: Now it is best to used when using the Zodiac,Ancient words,Ancient names, and the spelling of the number, such as Lotus,King,the zodiac Cancer..etc

Chapter 3

Etymology
The study of origins of words

What is etymology? Etymology is the study of the origin of words. The breakdown of the word etymology is as follows: Etymology: etym + olog + y
-Etym derives from *etymon*, a Classical Greek adverb that means **true, real, and actual.**
-+ ology (olog + y) derives from the Medieval Latin and Greek word *logia*, meaning **the study of.** It comes from the root of the Latin word *legein*, which means to speak (think of the word lecture).

Etymology is a a science that is very important, because the to world is fill **connotative meanings**, which I will explain. Meanings of a word has great power, so therefore we should want to know were the origin of the word comes from. The reason for this, if we don't know the **denotative meaning** of a word, when it comes to history, law, civics, and science, we can easily get lost in the concept of the meaning of the word.

<u>Denotation</u>-is the primary and literal meaning of the word,(*The Origin*)
<u>Connotation</u>-is the secondary meaning of the primary and literal meaning of the word *(not the origin)*

Now reading and researching words, is very important, so you will need to distinguish a noun from a adjective.

Noun- is a person,place, or thing
Adjective- is something describing a noun

*Now we need to understand the **root of the word**, and the **prefix** and **suffix**.*

Prefix-a word, letter, or number placed before another. **(which makes it the root of the word)**
Suffix-a morpheme added at the end of a word to form a derivative, **e.g.,*ation,-fy,-ing,itis.***

*Note the word can (**sound shift**), with added **vowels** and the **speech of a language**.
(Umlaut Linguistics) umlaut is a sound alteration,which is a vowel is pronnounced more like the following vowel or semi-vowel.

Linguistics is the scientific study of language and its structure, including the study of morphology, syntax, phonetics, and semantics. Specific branches of linguistics include sociolinguistics, dialectology, psycholinguistics, computational linguistics, historical-comparative linguistics, and applied linguistics.

Parts of Speech -Identifying the meaning of the word by knowing its function in a sentence (is it naming something like a *noun*? Is it stating an action like a *verb*? Is it modifying a noun like an *adjective*? Or does it modify a verb or adjective as an *adverb*?).

The words I will be using is in the European language.

The Based Languages
1.Greek
2.Latin
3.Germanic
4.English
And other Europeans Languages

So now let me show you some examples.

Light-*English* has two distinct words light. The one meaning '**illumination**' comes ultimately from Indo-European (root)**leuk-, louk-, luk-,** which also produced *Greek*- **leukos 'white'** (source of *English* -leukamia) and Latin- **lux 'light'** (from which *English* gets **lucifer**, literally meaning '**light-bearer**'), lumen '**light**' (hence *English* **luminous**), **lucere 'shine'** (source of *English* **lucid**), lustrare **'light up'** (hence English **illustrate** and **lustre**), and **luna 'moon'** (source of English lunar). Its main prehistoric *West Germanic* derivative was **leukhtam**, from which come *German* and *Dutch* **licht** and English **light**. The word lynx may be related.

Now **Greek** the oldest language in this based group, then **Latin, Germanic,** and **English**. So when any root word in coming in **Greek**, That will be the denotation meaning of the word. So if the The word Light, comes from the root word (**leuk-, louk-, luk-**) any word after that, using the same root, will literally mean **Light**.

*****Note**-As long as the word is pertaining to the root word denotation meaning, It still denotes.

Now take the word **(Demon)**

Demon-_English_ acquired this word from Latin in two forms, classical Latin **daemon** and _medieval Latin_ **demon**, which were once used fairly interchangeably for 'evil spirit' but have now split apart. Demon retains the sense **'evil spirit,'** but this was in fact a relatively late semantic development. **Greek -daimon** (source of **Latin -daemon**) meant **'divine power**, fate, god' (it is related to _Greek-_ **daiomai** 'distribute, allot,' which comes from an **Indo-European** base whose descendants include English tide and time). It was used in Greek myths as a term for a minor deity, and it was also applied to a **'guiding spirit'** (senses now usually denoted by **daemon** in English). It from this latter usage that the sense 'evil spirit'

Greek- **daiomai-** meant "divine power"
Latin- **daemon-** meant "divine power"
English-**daemon-** meant guiding spirit

We as people use a lot of words and their connotative meaning, so it is important to trace down that word origin, so we will not be lost in the words when we are dealing with history, law, and civics... etc. Cause if not, People in this world can tell you anything and you just follow that meaning.
Such as the word **Demon**, We believe the word demon means something **evil**, but when we use Etymology, when find out it means **Divine Power**.

*NOTE: This is the point we need to realize, research the origin and carefully read between the lines.
*Anytime you see the (Perhaps),(Maybe) and (Probably) that means the etymologist really wasn't sure of the meaning of that word, so we must dig deeper in history from other sources to find the facts.

Hypocrite- Etymologically, a hypocrite is someone who is 'playing a part,' merely pretending. The word comes via _Old French-_ **ypocrite** and late _Latin_ **hypocrita** from _Greek_ -**hupokrites** 'actor, hypocrite.' This was a derivative of **hupokrinein**, a compound verb formed from the prefix hupo- 'under' and **krinein** 'separate,' **which originally meant literally 'separate gradually,'** and eventually passed via 'answer' and 'answer one's fellow actor on stage' to **'play a part,'** and **hence 'pretend.'**

Greek- hupo-   meant "**under**"
         krinein- meant "**separate**"
Now when the word Compounds together,We get
HupoKrinein- originally meant literally '**separate gradually**,' In other words play a part or pretend

Now Here are some Indo Europeans words you can research.

Spirit: speis
Spring: sphereg
Mental: men
Passion: paen
Destiny: sta
Devil: guel
Free: prai
Bible: bibli
Belief: leubh
Belie: leugh

Chapter 4
Brief History of the Moors
"It is some references from different sources to research"

The History of the Moors/Muurs, date back over 10,000 years ago and more. The Moorish Flag is 10,000 years old, according to science it is 50,000 years old, and the red represents the bloodline of the Moorish people. In 1913 Noble Drew Ali, put the 5 point green star, which represents **"Love,Truth,Peace,Freedom and Justice".**

We can find this in **"101 Questions For Moorish Children" by** Noble Drew Ali
How old is the Moorish Flag ? – 50,000 years old (with math & science)

The Moors are who are the descendants of the ancient Moabites,Canaanites,Kemites,and other tribes out of Africa, They Had rule on earth for over 1,100 years. The ancient Moors navigated all across the world, and establish civilization. Which is known as **North West Amexem/North West Africa,**Which is the **"Moorocan Empire"** that consist of **Canada,America,Central America,South America, and the Joining Caribbean islands.**

In the "Circle 7"by Noble Drew Ali
The Moabites from the land of Moab who received permission from the Pharaohs of Egypt to settle and inhabit North-West Africa; they were the founders and are the true possessors of the present Moroccan Empire. With their Canaanite, Hittite, and Amorite bretheren who sojourned from the land of Canaan seeking new homes.

They migrated from northwest regions of Africa into Europe into the country Spain. They begin to conquer the regions in Europe, particularly Spain. They rule from (**711Ad to 1492 Ad**) The Moors were refer as the Moslems/Muslims within their Islamic Culture of science. When they had conquer Europe, in them times, they had came in contact with the Caucasian pale people, known to today as the so called white people. These Caucasian people was living in dark caves, crawling on all 4 legs,feeding off anything for survival. They were savage, and uncivilized people. They was genetically engineered and mutated from a monkey gene, created by **"Yakub"***Referred as Jacob in the holy bible,*which was a highly advanced scientist thousands years ago. It was the Moors who brought and taught the culture of masonry to civilize the pale people. The culture of masonry, deals with Astronomy and Geometry, which is the compass square used to build & measure things with precision. It was very necessary for masonry, to civilize the pale people, because it was a science survival kit, to understand time keeping for the sake of agriculture, so the pale people will understand how to harvest their crops to eat and feed their families.
The Caucasian people were dying off by the thousands, from the diseases and other infections,Until the Moors started to disinfect these people and civilize them on a path with a basic foundation of culture. Without the Moors civilizing the pale people, the

Caucasian group of people will be nearly or be extinct here today in the 21st century.

Here are some reference out the "Golden Age of the Moors" by Dr. Ivan Sertima As you see above that "711AD" is when the Moors had conquered Spain.

Below here, it speaks about that the moors expertise in medicine, and how they used masonry to build huge stones blocks, and was referred as the custodians of Ancient Kemetic Science.

Aside from comparable expertise in medicine and other areas of natural science which both the "medieval" Moors and ancient Kemites shared, I can also see a peculiar similarity in architectural skill. Dr. Ivan Van Sertima pointed out that at Lixus, Morocco and Gizeh in Egypt, there are the "finest African examples" of "fitted megalithic masonry."[115] We are told that the technique requires great skill, as the massive stone blocks which are fitted together are not of any one shape or size. Van Sertima says that the stones in these structures display "the complex regularity of patterns or designs in a jigsaw puzzle."[116] I find this to be a curious correlation between Morocco's Maures/Moors and Egypt's ancient Kemites, and I would relate this to observations and assertions of George James, regarding the Moors being "custodians" of ancient Egyptian erudition and culture.

Africans were pivotal also in the spread of Islam. The invasion of Spain in the eighth century and the survival of the Muslim dynasties in the eleventh owe a great deal to African military prowess and leadership. Chinyelu celebrates the military exploits of *Tarik* (who conquered Spain in 711 A.D.) of Yusuf Ibn *Tashifin*, leader of the Almoravides, who routed Alphonso VI's army in 1086 (15,000 Africans facing 70,000 Europeans) assuming leadership of Muslim Spain in 1091, and of *Yakub al-Mansur* who conquered Spain and Portugal on two separate occasions to become the most powerful ruler in the world. Such was the respect these leaders inspired in the hearts of their enemies, that royal crests and coats-of-arms in Europe were emblazoned with Moorish heads.

To the influence of Moorish science on Europe we finally turn, for it is in this field that the impact of the Moors is least known and most felt. Wayne Chandler points to advances in mathematics, the solving of quadratic equations and the development of new concepts of trigonometry. He informs us that Moorish chemistry refined upon gunpowder invention in China and thus introduced the first shooting mechanisms, known as firesticks. They were also known for their skill in medicine. For seven centuries the medical schools in Europe owed everything they knew to Moorish research. Vivisection as well as dissection of dead bodies was practiced in their anatomical schools and women as well as men were trained to perform delicate surgical operations. They were the first to trace "the curvilinear path of rays of light through air." This discovery in about 1100 AD is a prerequisite to the design of corrective eyeglasses. Students and teachers should read this essay also for its

In the book **"White Slavery in the Barbary States"** by Charles Sumner speak about that it was over 2 millions European Caucasians that were took as slaves after been capture by the Moors called the Barbary Coast Pirates. When the Europeans Caucasians start to rise in power in Europe, it was many wars between the Moors"Moslem" and The Caucasians "Christians" In **"1491"** which was the last strong hold for the Moors,That is when they had lost the war in Granada Spain, The Moors that was in battle at that time was refer as the "Turkish"Moors.

In the year **1491**, the Moors surrendered the city of Granada, their last stronghold. On November 25, **1491**, the Moors signed the **Act of Capitulation** after which Spain's King Ferdinand II and Queen Isabella took possession of the city. The Spanish Christians were **THANK-ful** for the Moors **GIVING (Thanks-Giving) up Granada,** ending Islamic rule in the country. This was a merry and festive day in the eyes of the Europeans and would henceforth be a day of THANKS. The defeat of the Moors brought 3 options: **(1) Replace their Islamic and Asiatic customs with European customs and Christianity, (2) Be expelled from Spain, (3) Face execution!** After the expulsion of the Moors, Cardinal Ximenes (backed by Pope Alexander VI of Rome and Queen Isabella) ordered the destruction of the beautiful and stunning Moorish libraries and mosques. Many Moors were driven into West Africa and were later captured and sold into America as slaves. Some of these slaves came on the slave ships of **Christopher Columbus (a Spanish Jew)** who was commissioned by Queen Isabella on his voyage to the New World. Many other Moors sought refuge in the country of Anatolia. These expelled Muslims[**Moors**] from Spain would later be called **Turks, (as in Turkish Moors)** and the country would later be renamed Turkey (by Europeans). After the expulsion of the Moors, Europeans began to annually celebrate November 25th. There would be great feasts all over Spain and Europe. Turkey bird was not the main meat dish served, but a fat roasted pig with an apple in its mouth (representing or symbolizing rebirth or a new beginning). Europeans ate so much on this feast day that virtually nobody could work the next day (which explains why in modern times many corporate jobs give their employees the day after Thanksgiving off!). Years later, Europeans began to take revenge on the Moors, now called Turks, because of the sexual relations the Moors held with white Christian Spaniard women. The black men [Moors] were mockingly called Turkeys, and were called this by European men because of their alleged wild and uncivilized nature? European Christians would now go on a Turkey shoot or Turkey hunt? The hunting of the stern Kharijite Moor from Turkey (Anatolia) became an act highly regarded by Europeans at the time. In the process of the turkey shoot Europeans would engage in the wholesale slaughter of the black-skinned Turkish Moors. Innocent Moors would be hunted down cold-blooded by European Christians. Many others would be captured, tied by their wrists and ankles with rope and roasted ALIVE (just like you roast that turkey bird in the oven for Thanksgiving dinner). The Moors that were not roasted alive were laid on the ground, and a European male would take a knife and literally CARVE up the Turk (which is what many people are reenacting when they carve the turkey bird on Thanksgiving). This bloody, carved up Turk **(Moorish man)** would then have his internal organs removed and his body stuffed with certain fabric material (the sick origin of stuffing the turkey bird with bread stuffing or dressing). The Moorish woman were literally carved and cut apart the way meat-eaters and butchers today cut up a chicken.

It was **January 1ˢᵗ,1492**, when the Europeans begin to celebrate the defeat of the Moors, which is celebrated as their **New Year** for the Europeans pale people not the Moors. The real new year is on **March 21ˢᵗ, The Spring/Vernal Equinox**. In other regions around the earth it is celebrated on the Summer Solstice.

APRIL FOOL DAY

In Spain, after centuries of Muslim rule, the Moors were overwhelmed by the Christian Army. However, the Moors were fortified in their homes. The Christian armies wanted to get rid of the Muslims somehow. They told the Muslims that they could leave their homes safely and could take only the necessary things from their homes. They were told they could sail away in the ships anchored on the quay side. The Muslims did wonder if this ploy was a trick. The Muslims were requested to go to the quay side to check the ships. They did so and were convinced. They then made preparations to leave. The next day, 1st April, they took their essential belongings and walked towards the quay side.

The Christians looted their homes and then set fire to them. Before the Muslims got to the ships, the Christians had set fire to the ships as well. The Christians then attacked the Muslims and killed them all:

Men, women and children. They then celebrated this carnage. This
then became a ritual that was celebrated every year and that has
been carried on until this day not only in Spain but in other countries too.

In 1492 ,Christopher Columbus which was a thief, murder, and crusader. He was
funded by queen Isabella and king Ferdinand on his voyage to
America"**Caribbeans Islands**". He had no knowledge on how to sail a ship and
he was told by the Portuguese that the people he will see, would look like the
(Indians)"Moors" that was still trap in Europe under the rule of European
Monarch. The Moors who stayed in Europe, they had their on agenda to, and they
agreed to help navigate the ships for Columbus. So when he came to the Americas,
Columbus intentionally called the people he seen Indians and stamp the word
Indian on the indigenous people here in America. It was 5 civilize Tribes of the
Moors.

1.El-Cherokee 2.Bey-Choctaw 3.Dey-Seminole 4.Al-Creek 5.Ali-Chickasaw

Here are five Kushite/Khemetian/Moorish names that ties Indians/Indo/India/Indios/Indus,
blacks to the sovernity of America: EL, Bey, Dey, AL, and Ali. These five names belong to
the Five Civilized Tribes. These tribes are EL - Cherokee, Bey - Choctaw, Dey - Seminole,
Al – Creek, and Ali – Chickasaw. The name El means God, force, or power. Bey means
ruler or land lord. Dey means knowledgeable. Al means the same as El, but Al (Goddess) is
feminine. Ali means the exalted or most high.

These Five Civilized Tribes came together to form a union called the Iroquois confederation.
This confederation formed the Articles of Confederation, which formed the American
government. The Articles of confederation became the United States Constitution, the
declaration of independence, and the Articles of Association. These documents are called
the four Constitutions.

El
Cosmos Law Giver
Legistlative
Branch

Bey
Governors of The
Land Enforce Laws of Eloheem
Executive Branch

It was over "**150 million**" people already in the Americas,So it make no sense when
European scholars speak of Columbus discovering something when people was already
there. It was a small percentage of "**Africans**"who came to America. They were trick by
their own people, the "**Dirty Moors**"who help the Europeans and enslave their own
people. It is Estimated around "**500,000**" and half of the population died from the bad
conditions on the ships, which will estimate around **250,000** people. So lets be clear, The
math does not even come close to adding up.
More info."**They were here before Columbus**"by Dr. Ivan Sertima

After 1492, The Europeans start to come in large numbers to America,But the Moors in

America wasn't allowing the Europeans to pass thru. The Moors were the true navigators, and patrol the seas. They also were ruthless and seized the ships of the Europeans.

They took over the hundreds of ships, demanded money and valuable resources, which became an annually tribute for the Europeans to pay every time if they wanted to pass threw. **Now Note: The Dirty Moors Navigated and Guided the ships. They knew that it was 3 currents within the sea,that will lead to north part of America,the Caribbeans, and central America**. In the time from the 1500s to the 1900s, it was many wars between the Moors and the European Christians. The Imperial Majesty of Morocco in(1786) decided to agree to a treaty with the Europeans Colonist to help stop the wars.

Morocco(America) is one of the first countries to recognize the independence of the **(United States)**as the "**Sultan Sidi Mohammed Ben Abdullah** "issued a declaration in **1777** allowing American ships access to Moroccan ports. In **1787** a Treaty of peace and friendship was signed in Marrakech and ratified in **1836**. It is still in force making it the longest unbroken treaty in the U.S history.
The U.S had also its first consulate in Tangier in 1797 in a building given by the sultan Moulay Sliman. It is the oldest U.S diplomatic property in the world.

.

Moroccan Treaty of Peace and Friendship between the United States and the Moors.

Treaty of Peace and Friendship, with additional article; also Ship-Signals Agreement. The treaty was sealed at Morocco with the seal of the Emperor of Morocco June 23, 1786 (25 Shaban, A. H. 1200), and delivered to Thomas Barclay, American Agent, June 28, 1786 (1 Ramadan, A. H. 1200). Original in Arabic.The additional article was signed and sealed at Morocco on behalf of Morocco July 15, 1786 (18 Ramadan, A. H. 1200). Original in Arabic. The Ship-Signals Agreement was signed at Morocco July 6, 1786 (9 Ramadan, A. H. 1200). Original in English. It is the longest unbroken treaty in the U.S history.

Now within that treaty, It gave the U.S. Colonist partial sovereignty within the Moorish Territory to do commerce, and business on the land of Morocco.If any person that had any contract with the U.S. and violated those terms,the U.S now had the authority to take legal action under the 1787 treaty of peace and friendship.

Al Moroccan(America) Is the sovereignty of the Territory or land of America.

(United States) are foreigners from Europe which has partial sovereignty to do business and commerce on the land of America.

Chapter 4
Law and Civics- *Claim your nationality*

We the people of African & Asiatic descent has been fool, condition, degraded, and separated from our fore fathers and fore mothers. When it comes to law and civics, We need to understand legal terms,contracts and etymology. It is certain code words and code names that the Europeans colonies has put in the minds of the indigenous people of this land here in America, such as **negro,color,black,Indian and African American thinking that is their identity of who they are"that is not a nationality,** We the Moorish Americas, are sovereignty by birthright and bloodline. Now, when I speak of the Americas,I am speaking of Canada,America,Mexico,South America, and the Caribbeans Islands. The Europeans are using these code words to suppress the sovereignty of the indigenous people of their land .It is so important to know your history and then you will know yourself.

Now on that Note, Study your history,law and civics. I quote, as the Native and ,indigenous people of the land of the Americas, I **"ENCOURAGE EVERY ONE OF AFRICAN and ASIATIC DESCENT TO CLAIM THEIR NATIONALITY"** It will put you back in your proper person as a human being and as being free under divine and common law.. So If you want to stop paying taxes, unnecessary fees, pressure to have driving license, when you have the right to travel upon your own land,...etc. **"CLAIM YOUR NATIONALITY"**

First of all, we have to make sure we know what a The word Moor means and what is it origin.

Many people believe that the word Moor means black. That would not make make any sense if Moor is used as a title as a Nationality, that refer you to a land,culture,civilization, and government by bloodline.. This can easily be proven, because at one point in time, we were called Blackamoors. And we as the native people will be calling ourselves "Black a Black" which make no sense, when it comes to nationality.

The word Black is an Adjective, which derives from the root word **"bhleg"**, meaning to shine and shimmer like a flame .The word **"Moor"** derives from the word **"Mer"** in Ancient Kemet, which meant many things, But the most common was overseer, chief officer,head, superintendent,director, foreman.

It also has meanings of collection of water, lake, pool, canal, flood, and stream.

So you will find in history of the Moors, they refer their self as the navigators of the sea, The Head, Chief of Government.

Research the book "Mer to Moor; Kemet Until now, by Cozmo EL

The concept of **"black"** as a metaphor for race was first used at the end of the 17th century when a French doctor named **Francois Bernier (1625-1688)**, an early proponent of scientific racism, divided up humanity based on facial appearance and body type." It is a **Adjective.**

To have a Nationality is will be a noun in the sense of a person,place or thing.

You cannot find any traces of our people calling themselves Black before **1500s**.

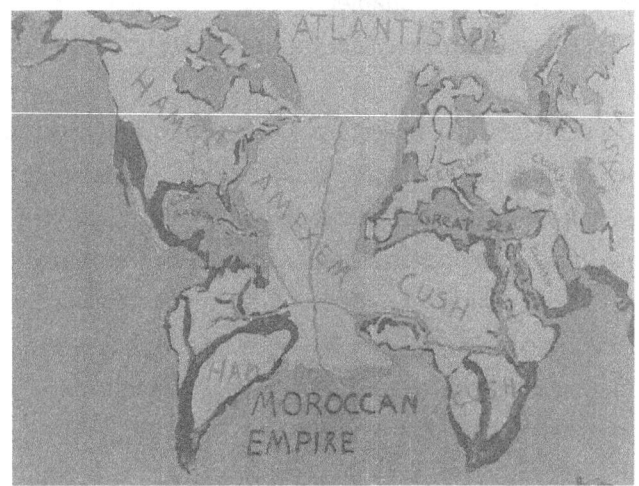

It is a global Nationality, meaning we were
Moors before the continents split up into the 7 continents we have now. So you have
Moorish Americans, Moorish Puerto Ricans, or Moorish Mexican, just like
how you can be from many different places (states,cities,counties etc..) in America, but
you are still an American.

*Their dominion and in habitation (the Moorish Empire) extended from North-East and South-West Africa, across
great Atlantis even unto the present North, South, and Central
America and also Mexico and the Atlantis Islands; before the great earthquake, which caused the great Atlantic
Ocean to sink under water over thousands years before B.C.E*

Birth certificate, social security number and driver license is a fraud and are instruments used by

the U.S. Government for human tracking. When your mother sign a birth certificate and sign off for a social

security number, they is selling their child into slavery under the jurisdiction of the United States Colonies. Now

when that happens, you as the native of your own land has gave up your sovereignty moral rights and became a

slave under color law, such as negro,color, and black which are not nationalities. So now you just have claim a legal

status, and became a legal residents a foreigner in your own land and you don't even know it, because we the moors

of African descent still want to believe that this is the so called white man land, and they got all the money and

control and it is nothing we can do but get a job and work for them and hope for the best. **"THAT IS A KOWARD

AND JUST PURE INGNORANT"** If you want to believe that and not research your ancestors history which has been

written by many African scholars such as **"Dr. Ben, Dr. John Henry Clarke, and Dr. Ivan Van Sertima"**, you will

forever be a slave to the false history that was brought by these modern day Europeans.

National Status vs Legal Status

National Status is that you is the sovereignty of this land here in the Americas by birthright and bloodline of the land that your fore-mothers ,and your forefathers descend from. You is not under the jurisdiction of the U.S. Corporation. You have rights and not privileges, which is under the U.S. Constitution.

Legal Status means that you is a legal resident, claiming negro,color, black or white and is a foreigner of the land in the Americas and you have no rights ,but only privileges, and is under the jurisdiction of the U.S. States as their property, The slave by contract.

That is in **(Title 18 U.S. Code of law)** you will see in legal terms that anything that is under the state is the possession of the United States.

So Now, **What is jurisdiction?** "Juris" meaning right, "Diction" meaning words or saying

Basically meaning they have the **"Right Words"** against you, and you don't, because you claim a **legal status** and gave up your **birthright**. Once we start to study our History from the Moorish scholars perspective and the law, these things will become more simple. Let's understand these concepts and principles.

Note: You is a corporation,property used as a slave by contract for the U.S. and you is representing that corporation as a fictional person, in what we called the strawman. A fictional person, that is just substance or has no substance, meaning you is just only a make believe person. So therefore, you do not have any rights under the U.S jurisdiction. So these things must be corrected on a National Standard.

•••The Contracts such as the Birth Certificate, Social Security, and Driver License all instruments used for the tracking of the U.S property. And by contract you is property and is plug into the matrix as a make believe person thinking it is a all real, "Property can't own Property" stop been slaves and wake the hell up.

Now, if you look on the back of your social security card .You can locate what Federal Reserve bank

owns you. The First letter in the back is how you will verify this.

(A) Boston
(B) New York
(C) Philadelphia
(D) Cleveland
(E) Richmond
(F) Atlanta
(G) Chicago
(H) St. Louis
(I) Minneapolis
(J) Kansas City
(K) Dallas

Driver License is a fraud and not your true identity, and it is only to be use for commercial use in driving vehicles
for the service of commerce under the state.
The U.S is using the **DMV as a agency** to steal your identity. Their job is to deprive you of your right to travel.
Remember, you is a corporation representing the state as a **"fictional person"** So now, how do this become legal
under law. Anytime your name is in all "caps" you is not in your proper person, so when that is so, the state can legal

hold you responsible of having a driver license to travel the highways because you is under their jurisdiction. "It is that simple"

Now here is some of the Supreme laws of the land that deal with the right to travel.

Personal Liberty:
The Right or Power of Locomotion; of changing one's
situation; or moving one's person to whatsoever place one's
own inclination may direct without imprisonment or
restraint unless by due course of law.

The Right of a citizen to Travel upon the public highways and to
transport one's property thereon, either by carriage or automobile, is not
a mere privilege which a city may prohibit or permit at will but a
common right which he / she has under the right to life, liberty, and the
pursuit of happiness. **Thompson v. Smith 154SE 579:**

"No state shall convert a liberty into a privilege, license it, and attach a
fee to it." **Murdock v. Penn., 319 US 105**

"If the state converts a liberty into a privilege, the citizen can engage in
the right with impunity." **Shuttlesworth v. Birmingham, 373 US 262**

"Traffic infractions are not a crime."
People v. Battle, 50 Cal. App. 3,step 1, 123 Cal.Rptr. 636,639.

"Speeding, driving without a license, wrong plates or no plates, no
registration, no tags, etc., have been held to be "non-arrestable offenses"
(Cal V. Farley, 98 Cal. Rep. 89, 20 CA 3d 1032.

The Right to Travel; The Right to Mode of Conveyance; The Right to
Locomotion are all absolute rights, and the police cannot make void the
exercise of Rights. *State v. Armstead, 60 s. 778, 779, and 781:*

This is all documented in Washington, D.C in the Library of Congress

What is a Policeman? They are **private security guards** and their job is not to enforce the law under the **Constitution**, their job is to **violate your rights,** so the state can gain profit and control of your movement, so by you not being in your **natural person** and **not knowing the law,** they come to intimidate you, so can waived your rights and they attach a fee to it and collect profits for the state. They are **Roman soldiers** under the order of **secret societies** which is the order of the **(pope of Rome and the knight of Columbus)** this is all part of the
king Alfred plan.

What is a Officer? A officer is a person who swears by oath to uphold the United States Constitution as
an Officer of the law of the Supreme Court decisions and the laws of the land and to protect the rights
of the Natural people and citizens of the land.

*Kolender v. Lawson (461 U.S. 352, 1983)*in which the United States Supreme Court ruled that a police officer could
<u>not</u> arrest a citizen merely for refusing to present identification.

YOU DO NOT TO HAVE REVEAL YOUR IDENTITY TO THE POLICE, BUT THEY DO.

YOU CAN SUE THE POLICE FOR AN ILLEGAL ARREST AND RESIST ARREST WITH IMPUNITY!

"An illegal arrest is an assault and battery. The person so attempted to be restrained of his liberty has the same right to use force in defending himself as he would in repelling any other assault and battery."

(State   v.   Robinson,   145   **ME.**   77,   72   **ATL.**   260).

Police Officers can only ask for your identification when an investigation is under way. And your part of it. Therefore, when they hinder you, they are saying that you are under **investigation**. Their car lights and sirens are to only go on if there is an **investigation or crime, and a traffic infraction is not a crime..** Therefore they must identify to you the investigation, and your part in it. This is why you ask them **"What is their probable cause".**

So if they don't have a probable cause, and they continue to detain you, that when you ask for their Supervisor or the Sheriff. And when that arrived to the scene, you ask them did they take the oath to support the U.S. Constitution, which is in "Article 6" if they can't present you no evidence of their oath and probable cause, and continue to detain you, that is a supreme violation of your rights. Now it becomes a law suite.

 Now, we can clearly see how they are confusing the public people with the law, because they know over half of the population do not know or just do not understand how to operate under those circumstances. The **U.S. Government and their Roman Soldiers all crooks**, and they will continue to do what they do, until the public people wake up and study **"what is law and what is not"**.

Note: All laws in America is base on the Constitution and it cannot be reinterpret no other way .

Ramesses

What is my status and nationality?

My Status, I am a natural born sovereign of the Moorish Empire, my Moorish ancestors have lived in this land for thousands of years. My nation Morocco is one of the first countries to recognize the independence of the United States of America and in 1787 a Treaty of peace and friendship was signed in Marrakech Morocco and ratified in 1836. It is still in force making it the longest unbroken treaty in the U.S.A. history.

Sovereign of the Moorish Empire

Treaty Peace Friendship

United Nations   United States of America

△ **Love – Peace – Truth – Freedom & Justice**
⊗ Kemetic Moorish American Culture --------- learn more at bemoor.com

Find your Nearest Moorish Science Temple

HOME OFFICE
2905 5th St. SE
Washington, D.C. 20032
(202) 547-7597

Moorish National Home NO. 1
Brother W Creighton-Bey, G.S.,
(703) 628-9380

Sister C. Tolbert-Bey, GS/MNH NO. 2
16910 Moorish Ave.
Bonner Springs, KS 66012
(913) 441-6745

Brother R. Hill-Bey, GS/Temple NO. 1
(773) 443-0688

Brother M. Hayes-Bey, GS/Temple NO. 4
15520 Harper Street
Detroit, MI 48224
(248) 881-8923

Brother T. Austin-El, GS/Temple NO. 5
2918 N Sarah Ave
St. Louis, MO 63115
(314) 669-9031
Email:mstasub5@yahoo.com

Sister D.Banks-Bey, GS/Temple NO. 7
131 Bartlett Ave.
Cleveland, OH 44120
(216) 470-4898

Brother R. Johnson-Bey, Jr., GS/Temple NO. 9
1000 N. Hoyne Ave.
Chicago, IL 60622

Brother W. Clendenin-Bey AGS/Temple NO. 10
call:(646) 372-7079

Brother A. Hopkins-Bey, GS/Temple NO. 11
2259 N. 5th Street
Philadelphia, PA 19133
(215) 939-3584

Know Thyself Show
Mondays 6:30PM EST.
Call in number 646-478-4713

Brother D. A. Siggers-Bey, GS/Temple NO. 12
1200 W Sugar Creek Rd
Room 200
Charlotte, NC 28213
336-483-6035
Mailing Address:
P. O. Box 790601
Charlotte, NC 28206
Email:mstatemple12@yahoo.com

Brother G. Gregory Bey,
Grand Governor Of Alabama
Branch Temple NO. 12
P.O. Box 610522
Birmingham, Alabama 35261
Email:gregbey1971@gmail.com
(205) 810-5267

Sister S. Nunlee-Bey, GS/Temple NO. 13
P.O. Box 430803
Pontiac, MI 48343-0803
(248) 759-9834

Brother A. Frazier-Bey, GS/Temple NO. 15
328 Walnut St
Hope, IN 47246-1559
(812) 546-5281

Brother P. Kennedy-Bey, GS/Temple NO. 18
316 N. Michigan Street Suite 111
Toledo, OH 43604
(419) 340-8682
Email:pbey@bex.net

Brother O. Brown-El, GS/Temple NO. 19
4310 Martin Luther King blvd
Flint, MI 48505
(810) 785-8544

Brother C. Gray-El, GS/Temple NO. 20
394 Ferry Street
Pontiac, MI 48342
(313) 772-1601

Brother N. Revels-Bey, GS/Temple NO. 21

349 Bainbridge Street
Brooklyn, NY 11233-1903
(718) 452-1015
Email:mstany21@gmail.com
moorishsciencetempleofamericainc.org
Facebook Page

Brother C. Fuqua-Bey, GS/Temple NO. 25
5601 Grand River
Detroit, MI 48208
(313) 717-3670
Email:grandgovernormav@gmail.com

Sister H. Graves-El, GS/Temple NO. 28
Marian & Stanger Ave
Glassboro, NJ. 08028
(856) 696-2411/(856) 881-8404

Brother R. Dexter-El, GG
Moorish American Study Group AZ, NV and CA
AZ (480) 639-4586,
NV (702) 609-8334,
Cell (313) 779-6917

Brother J. Young-El, GS/Temple NO. 33
1932 Brown Street
Philadelphia, PA 19130
(215) 765-7647

Brother Y. Sirius-El, GS/Temple NO. 34
237-239 Hancock St.
Brooklyn, NY 11216
(718) 300-8171
Facebook Page

Brother R. Hadley-El, GS/Temple NO. 43
33540 Lipke Street
Clinton Township, MI 48035
(313) 948-8868

Sister R. Higgins-Bey, GS/BT NO. 43
Brother M. Cook Bey, G.G.,
Sister S. Lee Bey, D.M.,
P.O. Box 80187
Milwaukee, WI 53208-9998
(773) 712-2975
(414) 514-9821
(414) 208-0115

Brother J. Williams-Bey III, GS/Temple NO. 47
2731 Beacon Hill Ct.
Wichita, KS 67220
(316) 687-2105

Brother J.Florence-El, GS/Temple NO. 48
671 Pennington Ave.
Trenton, NJ 08629
(609) 954-4572

Brother D. Clark-El, GS/Temple NO. 54
2230 Light Street
Bronx, NY 10466
(646) 623-0929
Email:GrandSheik54@aol.com

Brother E. Johnson-Bey, GS/Temple NO. 55
4182 N. 700 E.
Hope, IN 47246
(812) 546-5242

Sister D. Warner-Bey, GS/Temple NO. 57
105 Sheffield Ave.
Lockport, IL 60441
(815) 726-0378
(cell) 815-726-0378
Historical Society Of Islamism

Brother J. Fielder-Bey, GS/Temple NO. 65
1700 N 18th Street
Kansas City, MO 66102
(816) 564-6496Email:

Brother P. Chase-El, GS/Temple NO. 71
732 Webster Street, NW
Washington, D.C. 20011
(202) 726-5025
Email:temple71mstofa@verizon.net

Brother L. Lee-Bey, GS/Temple NO. 75
P.O. Box 771488
St. Louis, MO 63107
(314)374-5141

Brother M. Moses-El, GS/Temple NO. 77
2905 5th St. SE
Washington, D.C. 20032
(301) 642-4235

Brother W. Barnes-El, GS/Temple NO. 78
5248 Reisterstown Road
Baltimore, MD 21215
(443) 540-5546

Brother A. Slaughter-El, GS/Temple NO. 88
1083 Austin Ave. N.E.
ROOM 5
Atlanta, GA 30307
(770) 855-7494
Mailing Address:P.O. Box 962271
Riverdale GA. 30296
Email:anville@comcast.net
Facebook Page

Brother O.German-Bey,G.G. of Georgia
Email:ogermanbey@gmail.com

Study Group of Wichita, Kansas
1123 N Wabash Ave.
Wichita, Kansas 67214
(316) 371-3706
Email:mcgheebey@hotmail.com
Brother S. McGhee Bey, Grand Governor of Kansas

Weekly Conference Call
Sunday 7:30pm/EST, Koran Class
Thursday's Questions and Answers
7:30pm/EST
Call in Number 1-712-432-3447 code 699913#.

West Coast Conference Call:
AZ, NV and CA,
call for information,
313-779 6917 or 602-814-3600
(Mountain standard time same as PST)
Friday: 7:30 pm to 9:30 pm
Sunday: 3:00 pm to 5:00 pm
Wednesday: 7:30 pm to 9:30.

Every Wednesday's Conference Call:
Call in Number 1-712-432-6333 code 312751#
Time: 7:00PM EST 6:00PM Central Time
Facilitators:
Brother J. Fuqua Bey and Sister Porter El

Moorish Science Temple of America in Durham   Contact Phone

902 Old Fayetteville St   P.(919) 491-5182

Durham , NC27701

United States

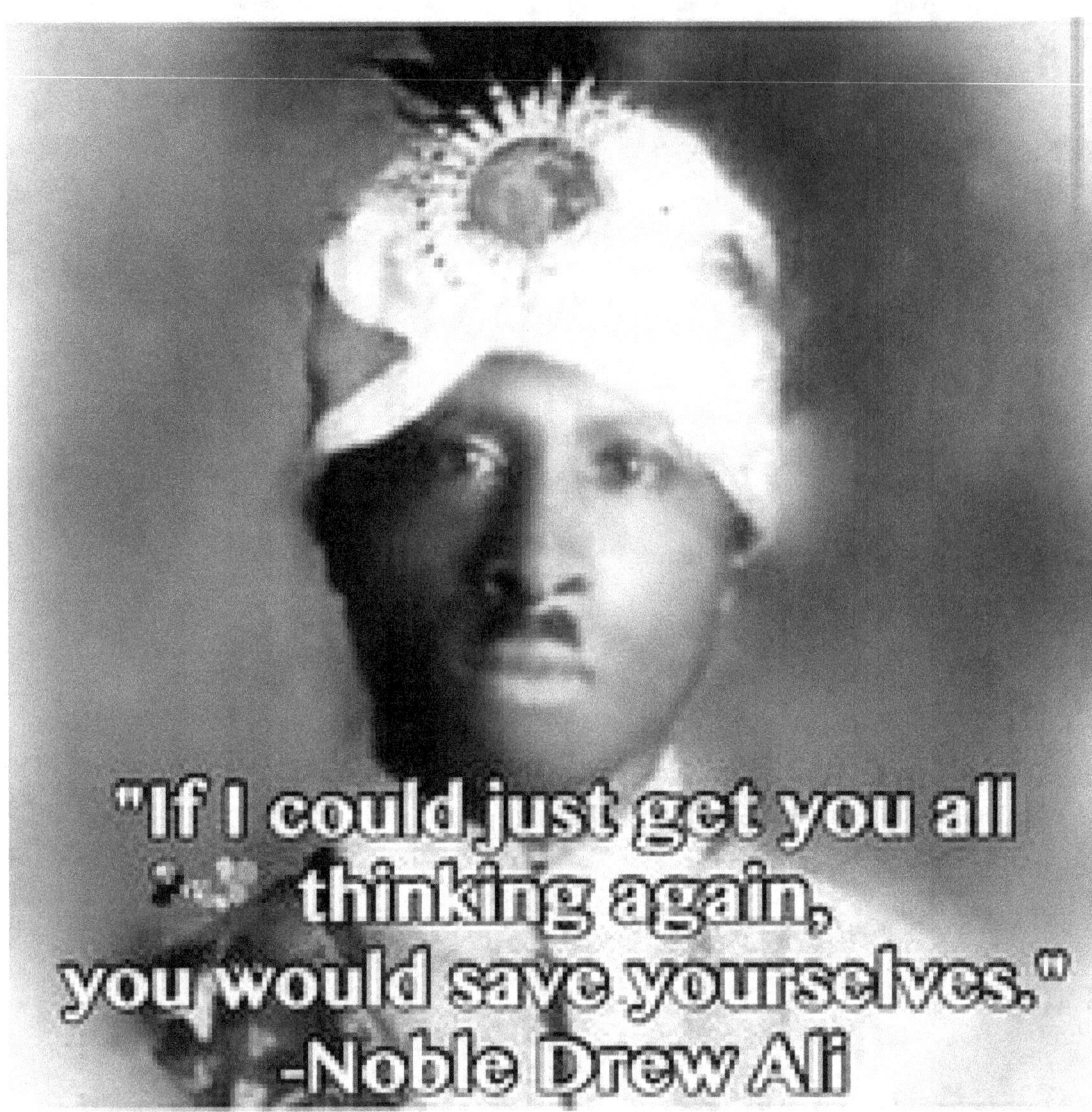

"If I could just get you all thinking again, you would save yourselves."
-Noble Drew Ali

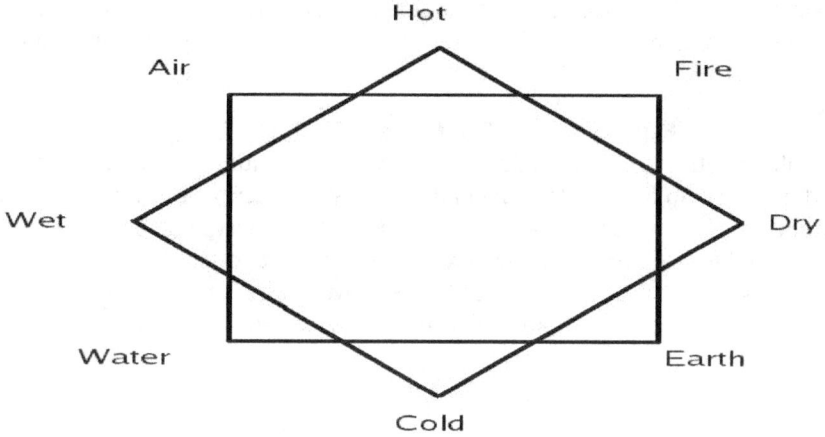

These are the 4 Elements which connect the nature of each zodiac.
"The idea is to know your zodiac sign to know and understand the nature of self on cosmic level."

Earth-Relaxed and Solitary
Water-Social and Relaxed
Air- Energetic and Social
Fire-Energetic and Solitary

Hot-Energetic
Wet-Social Solitary
Cold-Relaxed
Dry-Solitary

Earth-Knowledge and Learning
Water-Understanding and Assimilation
Air-Communication and Communication
Fire-Thinking and Expressive Thoughts

Fire Signs: Aries,Leo,Sagittarius
Earth Signs: Taurus, Virgo, Capricorn
Air Signs: Gemini, Libra, Aquarius
Water Signs: Cancer, Scorpio, Pisces

Now to have harmonious balance with compatibility, each element must attract according to the law of nature.
For example, **(Water)** is **Relax/Calm"Wet and Cold"** and **(Fire)** is **Energetic."Hot and Dry"**
Water can help calm down the high energetic nature and create Harmonious Balance between the two.
Also **(Water)** is Understanding **(Fire)** Expressive Thoughts, and water take the time to listen to fire expressive thinking or thoughts of ones situation.
Now **(Earth)** is **Relax and Solitary "Cold and Dry"** and **(Air)** is Energetic and Social,**"Hot and Wet"** **(Earth)** is not social,and is to self, **(Air)** is social and more communication. They will find many ways to communicate to create balance.

Earth Signs:Taurus, Virgo, Capricorn

Earth people are solitary,grounded, practical, disciplined and focused. They people are such perfectionists, they are always ready to take on any job or task themselves just to ensure it's done properly. Even then, earth people don't feel like they or anything they do is ever good enough. They are those who is willing to learn new things.

Water Signs:Cancer, Scorpio, Pisces

Water people are social,calm,emotional, intuitive, deeply creative, and spiritual. Water allows people to emotionally connect with others and find ways to fully understand things. And yet, water people are so sensitive that they often have a hard time unplugging from life's chaos. Water people tend to be secretive and private. It is important then, for water people to learn how to balance their emotions through meditation and see the value in water. Water is real. Water makes us vulnerable and it makes us human. It connects us with spirit and with soul. If you want to achieve a happiness and balanced, you need water.

Air Signs:Gemini, Libra, Aquarius

Air people are very social,brilliant, curious, independent, talkative (they literally fill the air with words and more words), observant, and entertaining, but they are also impractical and restless. Air people are intellectuals always on a quest for new information. So air people have an enormous ways to develop new ideas and tell stories. They have a hard time truly connecting emotionally to others, even though they want nothing more than to be completely understood. It's Critical for air people to learn how to give weight to their words and tap into the earth and so they will be grounded.

Fire Signs:Aries, Leo, Sagittarius

Fire people are highly energetic ,enthusiastic, inspirational, funny, dramatic and fun. They are natural performers. While fire people easily swing from one extreme to the other, it is important to remember that fire people speak and act straight from the heart or sometimes from their emotions or the sub-consciousness They deliver everything with a lot of passion. Fire people easily grow self-conscious when they speak before thinking (which happens fairly often). So people of fire, must learn to communicate ,and not to be to pushy out of a desire to help. The fire person's challenge is to learn to tame the "fiery beast" inside and create balance by drawing from the three other elements: water, earth and air.

Note:Understand the nature within your element and zodiac sign, but it is times you will have to step out your nature to create balance in certain situations. Each element builds and connect with others in many ways in the universe.

"If you want to understand the Universe, you must study thy self".

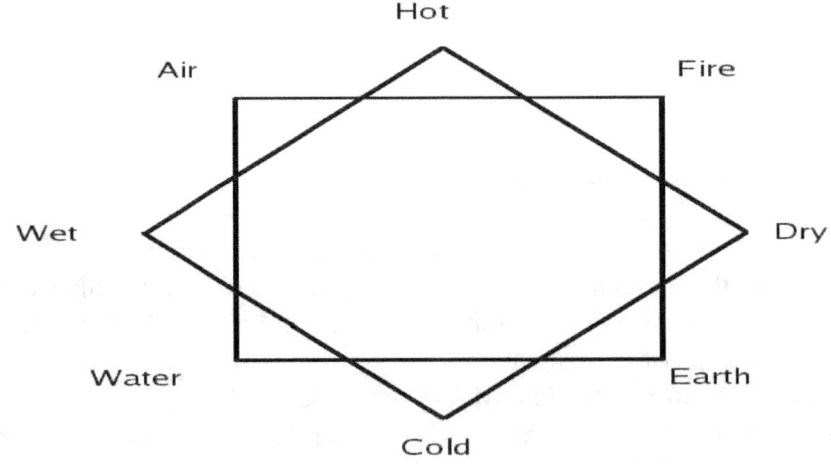

fire   earth   air   water

aries   taurus   gemini   cancer

leo   virgo   libra   scorpio

sagittarius   capricorn   aquarius   pisces

Chapter 7

"The 7 Chakras, Music and Vibration"

What is the science behind the **7 chakras** and the **kundalini energy?**
The word **"chakra"** meaning is **circle, vortex, wheel and center**. It is **88,000 chakras** in the human body, and have **7 major energy points** that are center in the middle of the human body. These energy points have major effects on how we humans move and think in every day life. When these energy points are not balance, It will cause the body to be **dysfunctional and imbalanced** by putting stress on certain organs, muscles and nerves. Now, when these energy points are in **correct balance** and activated, the mind and body begin to become in tune with nature. **Consciousness is the key.**

Now, What is the **kundalini energy?** In **Sanskrit** it means **snake or serpent,** meaning that it **coils or spirals.** It is the energy that start from the **base of the spine (Root Chakra) up to the pineal gland (Crown Chakra).** So when the **7 chakras** is stimulated or active, it coils and spirals like a snake to activate the higher brain senses. **(Your consciousness)** Now your when your higher self is awake and you have **"Light"** with the power to **transmute and transform energy to different forms. "That is Alchemy"** So on that note, that is what makes a person **"Illuminati or Illuminated"** because now that have light. So do not let that word **Illuminati** in the **secret societies** confuse you in thinking its is something bad, it is a word that comes in Latin, meaning light, so if I have light, **"I am Illuminati"**

The the first **Three Chakras** are already activated, which is the **1.Root Chakra,** is survival instinct **2.Sacral Chakra** is emotions , and **3. The Solar Plexus or Navel Chakra** is will power. These three lower chakras deal with the **Physical World** and the key is to balance them out. The higher chakras is **4. Heart Chakra** it is love and connection, **5.Throat Chakra** communication is, **6.Brow Chakra** intuition **7.Crown Chakra** is Enlightenment, being one with the universe. The key is to activate these chakras threw meditation, breathing techniques, music **(432hz)** frequency, Healing crystals, metals, beads, herbs and a alkaline diet. These higher chakras are unseen energy which deals with the **Spiritual World.**
"Balance is key with all 7 energy points"

Note: Consciousness is your money, and you have the power to determine your wealth in life.

Now, here is a chart breaking down the power functions of each chakra.
(Study this chart)

Chakra	1. Base	2. Sacral	3. Solar Plexus	4. Heart	5. Throat	6. Brow	7. Crown
Symbol	Lotus Petals: 4	Lotus Petals: 6	Lotus Petals: 10	Lotus Petals: 12	Lotus Petals: 16	Lotus Petals: 96 (48 X 2)	Lotus Petals: 1,000
Sanskrit Name	Muladhara	Svadhisthana	Manipura	Anahata	Vishuddah	Anja	Sahasrara
Pronounciation	moo-*lah-dar*-uh	swa-deesh-*stahn*-uh	manee-*pour*-uh	a-*nahh*-hat-tuh	vee-*shoo*-duh	*awdg*-n'yah	sahas-*rar*-uh
Translation	Root	Sweetness	Lustrous Gem	Unstruck	Purification	To Perceive	Thousand-fold
Center of	Survival	Emotions & sexuality	Will Power	Love & Connection	Communication	Intuition To see	Enlightenment & Spirituality
Location	Base of spine	Pelvis	Belly (between solar plexus and navel)	Heart	Throat	Behind forehead	Top of head
Color	Red	Orange	Yellow	Green (or Pink)	Sky Blue	Indigo Blue	Violet (or White)
Musical Note	C	D	E	F	G	A	B
Sound	Lum	Vum	Rum	Yum	Hum	Aum	Silence
Element	Earth	Water	Fire	Air	Sound	Light	Thought
Planet	Saturn	Moon	Mars	Venus	Mercury	Jupiter	Uranus
Gland	Adrenals	Reproductive	Pancreas	Thymus	Thyroid	Pituitary	Pineal
Sense	Smell	Taste	Sight	Touch	Hearing	Intuition	n/a
Right(s)	To Be, To Have	To Feel, To Want	To Act	To Love, Be Loved	To Speak, Be Heard	To Perceive	To Know
Gemstones	Red or Black Garnet, Red	Carnelian, Sunstone, Peach Moonstone,	Golden Tiger's Eye, Citrine,	Malachite, Green Aventurine, Emerald,Gree	Angelite, Aquamarine, Turquoise, Blue Lace	Lapis Lazuli, Amethyst, Iolite,	Amethyst, Lepidolite, Charolite, Clear Quartz,

Jasper, Ruby, Hematite, Obsidians, Smoky Quartz, Bloodstone	Red, Peach, and Orange Aventurine	Amber, Yellow Jasper, Honey Calcite, Golden Jade	n Jade, Peridot, Rose Quartz Rhodochrosite , Rhodonite	Agate, Amazonite	Sodalite, Charoite, Sugilite, Blue Sapphire	White Opal, Selenite, Rainbow Moonstone, Diamond

Remember, Every 7 major energy point is critical, so "Balance" will always be the key.

To activate the the higher energy points you must practice breathing techniques.

Balance Breathing Technique

Close the eyelids. Press the eyes up gently and focus at the brow point(the top of the nose where the eyebrows meet.

1)Left Nostril Breathing. Sit in a Easy Pose. Rest the left hand on your thigh,touch the tip of the thumb with the tip of the index finger).The left arm is straight on the left knee. Raise the right hand in front of the face with the palm flat facing to the left. The fingers of the hand are together and point straight up. Press the side of the thumb on the right nostril to gently close it. Begin long,deep, complete yogic breaths through the left nostril. Inhale and exhale only through the left nostril. Continue for 3 minutes. Inhale and hold comfortably for 10-30 seconds exhale and relax.

2)Right Nostril Breathing will be the opposite of the left nostril breathing exercise.

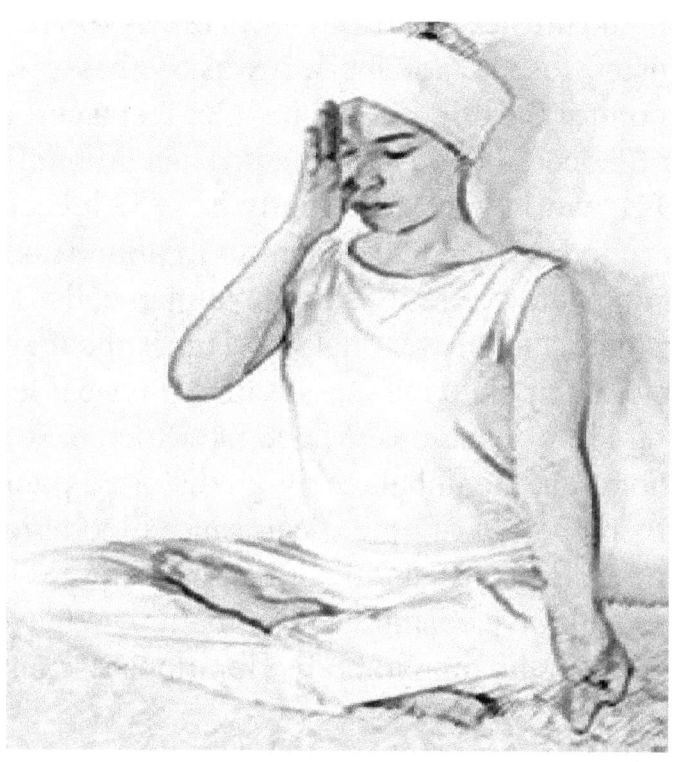

"Music and Vibration"

(432 hz)

What is 432hz? It is a harmonic of light,frequency, that resonates to human cells, which unifies space,light,time,matter,gravity and electromagnetism.

What is Hertz/Hz? It is a unit of measurement of vibrations and waves.

What is vibration?It is a periodic back-and-forth motion of the particles of an elastic body or medium, commonly resulting when almost any physical system is displaced from its equilibrium condition and allowed to respond to the forces that tend to restore equilibrium.

What is Frequency?The rate at which a vibration occurs that constitutes a wave, either in a material (as in sound waves), or in an electromagnetic field (as in radio waves and light), usually measured per second.

The Europeans change the music scale from 432 Hz to 440 Hz, It was dictated by Nazi propaganda minister, Joseph Goebbels. It was used as a war tactic against the enemy. He used it to control the mind of the people, feel a certain mood, and to make them a prisoner of a certain consciousness. Then around 1940s the United States introduced 440 Hz worldwide, and finally in 1953 it became the ISO 16- standard. That is why in today music you hear, it vibration is on a 440hz d frequency, which is a unnatural vibrations to the human cells. The 440hz are low vibrations, which stimulate the 3 lower chakras. This is the main reason the music has been dumb down to manipulate the masses on a physical level to keep people far way from becoming more conscious. Music like Rock & Roll,dumb down rap music is a very low vibration, the unbalance high distorted pitches, is detrimental to the human body. Over a period of time it can cause blood pressure to rise,mental stress and many other problems. Music in 432hz, is a natural vibration that we as human is in tune with on a cosmic level of nature. When in tune with the 432hz frequency, the cells in the human body vibrations rise, and automatically starts to heals itself.

Note:You can convert your music to 432hz with audacity for your personal used.

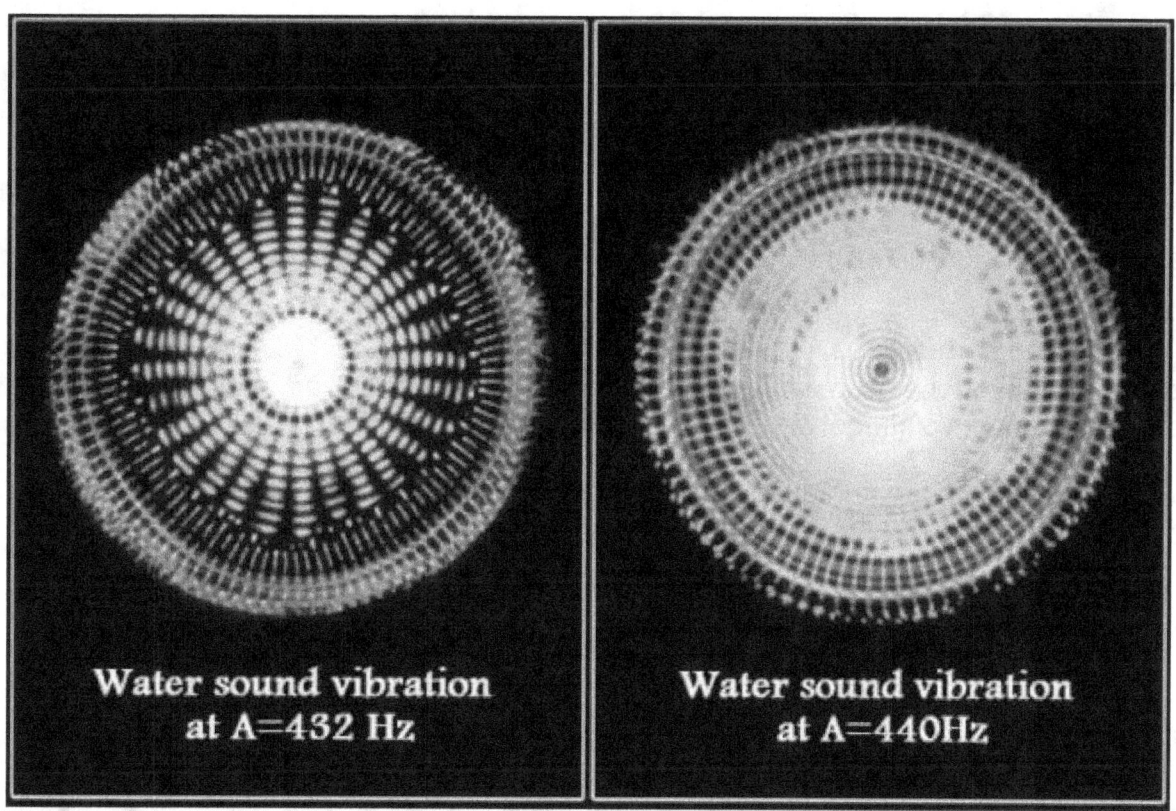

Water sound vibration
at A=432 Hz

Water sound vibration
at A=440Hz

Here in this chart below is showing the harmonics and overtone. When looking at the frequency, the wave pattern is form the same way as the human DNA.

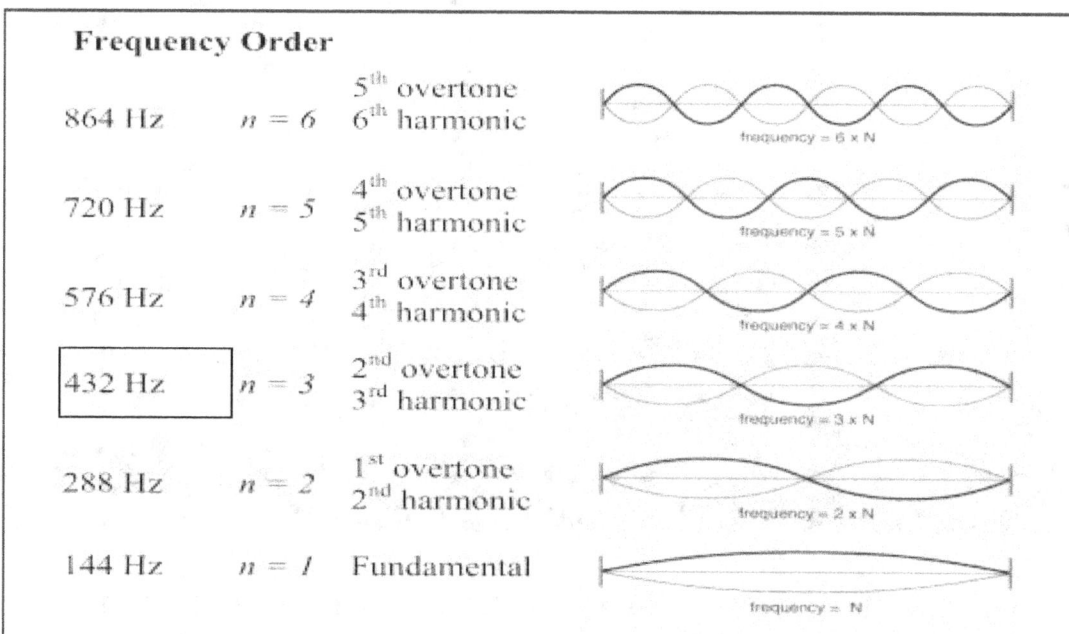

Frequency Order

864 Hz	$n = 6$	5th overtone 6th harmonic	frequency = 6 x N
720 Hz	$n = 5$	4th overtone 5th harmonic	frequency = 5 x N
576 Hz	$n = 4$	3rd overtone 4th harmonic	frequency = 4 x N
432 Hz	$n = 3$	2nd overtone 3rd harmonic	frequency = 3 x N
288 Hz	$n = 2$	1st overtone 2nd harmonic	frequency = 2 x N
144 Hz	$n = 1$	Fundamental	frequency = N

The earth has 33 Harmonics and 32 overtones. These Harmonics and Overtones is a key thing when dealing with music,vibration, and sound. That is why the number 33 is so significant in Masonry. It also connects to the 33 vertebrates in the human body of the rising of the kundalini energy. The Earth Vibrates at rate of 7.83,which is natural vibration on a cosmic scale. To understand more, we must know the meaning of a harmonic and overtone.

What is a Harmonic?
(Music) It is an overtone accompanying a fundamental tone at a fixed interval, produced by vibration of a string, column of air, etc., in an exact fraction of its length.

(Physics) It is a component frequency of an oscillation or wave.

What is a Overtone? A musical tone that is a part of the harmonic series above a fundamental note and may be heard with it.

(Physics) a component of any oscillation whose frequency is an integral multiple of the fundamental frequency

THE CHAKRA TONES BASED ON THE 432 HZ GRID

1ST CHAKRA ROOT	2ND CHAKRA SACRAL	3RD CHAKRA SOLAR PLEXUS	4TH CHAKRA HEART	5TH CHAKRA THROAT	6TH CHAKRA 3RD EYE	7TH CHAKRA CROWN
Mulakhara	Swadhisthana	Manipura	Anahata	Vishudhi	Adzhna	Sahasrara
GROUNDED	OPEN	CONFIDENT	COMPASSIONATE	EXPRESSIVE	INTUITIVE	CONNECTED
NOTE C	NOTE D	NOTE E	NOTE F#	NOTE G	NOTE A	NOTE B
TONES:	TONES:	TONES:	TONES:	TONES:	TONES:	TONES:
128 HZ	144 HZ	162 HZ	182.25 HZ	192 HZ	216 HZ	243 HZ
256 HZ	288 HZ	324 HZ	364.5 HZ	384 HZ	432 HZ	486 HZ
512 HZ	576 HZ	648 HZ	729 HZ	768 HZ	864 HZ	972 HZ

Note:You will like make decisions on a positive vibration that you feel comfortably with, if the vibration is negative, It is best to fall back if you is not sure of the results or outcome."Listen to what your body is telling you."

Chapter 8
Warfare Against the People

This chapter is a reference base were you can research on your own and analyze the information for self.

Their base strategy is to create the **"Problem"** see the **"Reaction"** and have the **"Solution"** for it.
"Problem, Reaction & Solution"

The King Alfred Plan
"Rex 84"

In the event of widespread, continuing and coordinated racial or civil disturbances in the United States of America, KING ALFRED, at the behest and discretion of the President, is to be put into action immediately.

Participating Federal Agencies
National Security Counsel (NSC)
Department of Justice (DOJ)
Central Intelligence Agency (CIA)
Department of Defense (DOD)
Federal Bureau of Investigation (FBI)
Department of Interior (DOI)
Federal Emergency Mgmt. Agency (FEMA)
Immigration Naturalization Service (INS)

Participating State Agencies
(Under Federal Jurisdiction)
National Guard Units
Reserve Units of the Army
Highway Patrol Agencies
State Police Agencies
State Emergency Mgmt. Agency (SEMA)
Disaster Relief & Civil Disturbance

Participating County & Local Agencies
(Under Federal Jurisdiction)
County Police
Local Town & City Police

Even before 1954, when the Supreme Court of the United States of America declared unconstitutional separate educational and recreational facilities, racial unrest and discord had become very nearly a part of the American way of life. But that way of life was very repugnant to most Americans. Since 1954, however, that unrest has resulted in the loss of life, limb, and property, and has cost the taxpayers of this nation tens of billions of dollars. And the end is not yet in sight. With Law Enforcement Agencies in this country continuing to exercise excessive force and murder of its So-called African-American citizenry. The statistics indicate that a violent confrontation is on the horizons. Why in fact given the recent uprisings in St. Louis and Cincinnati, it appears inevitable. This same injustice and violence has raised the tremendously grave question as to whether the races can ever live in peace with each other.

Each passing year and month has brought new intelligence that, despite new laws, programs passed or monies to alleviate the condition of the so-called Minority, the so-called Minority (African-Americans) is still not satisfied. Demonstration, rallies, sit-ins, marching and rioting have become a part of the familiar scene. Federal troops, State Police and National Guard Units have been called out in city after

1

city across the land, and America\'s image as a world leader is severely damaged. Our enemies (U.S.A.\'s)

Press closer seeking the advantage, possibly at a time during one of these outbreaks of violence. The Minority has adopted an almost military posture to gain its objectives, which are not clear to most Americans. It is expected, therefore, that, when those objectives are denied the Minority (So-Called African-Americans), racial war must be considered inevitable. When that emergency comes, we must expect the total involvement of all 34 million members of the Minority, men, women and children, for once this project is launched, its goal is to terminate, once and for all, the Minority threat to the whole of the American society, and indeed, the Free World.

Preliminary Memo: Department of Interior

Under KING ALFRED the nation has been divided into 10 Regions. In case of a National Emergency or Civil Disturbance, Minority members will be evacuated from the cities by federalized national guard units, federal troops and state and local police agencies which have become federalized under Marshall Law or Executive Order. The Minority members will be transported using public, commercial or military transportation, and detained in nearby military interment centers, prisons or military installations units until a further course of action has been decided.')

F. E. M. A. (Federal Emergency Management Agency)

The FEMA Gulag: SECRET CONCENTRATION CAMPS

The September issue of THE OSTRICH reprinted a story from the CBA BULLETIN which listed the following principal civilian concentration camps established in GULAG USA under the \"Rex \'84\" program: Ft. Chaffee, Arkansas; Ft. Drum, New York; Ft. Indian Gap, Pennsylvania; Camp A. P. Hill, Virginia; Oakdale, California; Eglin Air Force Base, Florida; Vandenberg AFB, California; Ft. Mc Coy, Wisconsin; Ft. Benning, Georgia; Ft. Huachuca, Arizona; Camp Krome, Florida. The February OSTRICH printed a map of the expanding Gulag. Although this listing and map stirred considerable interest, the report was not new.

For at least 20 years, knowledgeable Patriots have been warning of these sinister plots to incarcerate dissidents opposing plans of the \"Elitist Syndicate\" for a totalitarian \"New World Order\". Indeed, the plot was recognized with the insidious encroachment of \"regionalism\" back in the 1960\'s. As early as 1968, the \"greatest land steal in history\" leading to global corporate socialism, was in a \"\"Master Land Plan\"\" for the United States by \"Executive Orders\" involving water resource regions, population movement and control, pollution control, zoning and land use, navigation and environmental bills, etc. Indeed, the real undercover aim of the so-called \"Environmental Rennaissance\" has been the abolition of private property.

All prelude to the total grab of the \"World Conservation Bank\", as THE OSTRICH has been reporting. The map on this page and the list of executive orders available for imposition of an \"emergency\" are from 1970s files of the late Gen. \"P. A. Del Valle\'s\" ALERT, sent us by \"Merritt Newby\", editor of the now defunct AMERICAN CHALLENGE.

Wake up Americans!\" The Bushoviks have approved Gorbachev\'s imposition of \"Emergency\" to suppress unrest. Henry Kissinger and his clients hardly missed a day\'s profits in their deals with the

2

butchers of Tienanmen Square. Are you next?

SUBJECT: Executive Orders APPLICABLE EXECUTIVE ORDERS

The following \"Executive Orders\", now recorded in the Federal Register, and therefore accepted by Congress as the law of the land, can be put into effect at any time an emergency is declared:

10995 – All communications media seized by the Federal Government.

10997 – Seizure of all electrical power, fuels, including gasoline and minerals.

10998 – Seizure of all food resources, farms and farm equipment.

10999 – Seizure of all kinds of transportation, including your personal car, and control of all highways and seaports.

11000 – Seizure of all civilians for work under Federal supervision.

11001 – Federal takeover of all health, education and welfare.

11002 – Postmaster General empowered to register every man, woman and child in the U.S.A.

11003 – Seizure of all aircraft and airports by the Federal Government.

11004 – Housing and Finance authority may shift population from one locality to another. Complete integration.

11005 – Seizure of railroads, inland waterways, and storage facilities.

11051 – The Director of the Office of Emergency Planning authorized to put Executive Orders into effect in \"times of increased international tension or financial crisis\". He is also to perform such additional functions as the President may direct.

A Dangerous Fact Not Generally Known

THESE EXECUTIVE ORDERS GROSSLY AND FLAGRANTLY VIOLATE ARTICLE 4 SECTION 4 OF THE CONSTITUTION OF THE UNITED STATES. \"THE UNITED STATES SHALL GUARANTEE TO EVERY STATE IN THIS UNION A REPUBLICAN FORM OF GOVERNMENT, AND SHALLPROTECT EACH OF THEM AGAINST INVASION; AND ON APPLICATION OF THE LEGISLATURE, OR OF THE EXECUTIVE (WHEN THE LEGISLATURE CANNOT BE CONVENED) AGAINST DOMESTIC VIOLENCE.\" \"REGIONAL GOVERNMENT IS NOT A REPRESENTATIVE REPUBLICAN FORM OF GOVERNMENT!\"

When Government gets out of hand and can no longer be controlled by the people, short of violent overthrow as in 1776, there are two sources of power which are used by the dictatorial government to

3

keep the people in line: the Police Power and the Power of the Purse (through which the necessities of life can be withheld). And both of these powers are no longer balanced between the three Federal Branches, and between the Federal and the State and local Governments. These powers have been taken over, with the permission of the Federal Legislature and the State Governments, by the Executive Branch of the Federal Government and all attempts to reclaim that lost power have been defeated. Stated simply: the dictatorial power of the Executive rests primarily on three basis: Executive Order 11490, Executive Order 11647, and System which is operated through the new and all-powerful Office of Management and Budget.

E. O. 11490 is a compilation of some 23 previous Executive Orders, signed by Nixon on Oct. 28, 1969, and outlining emergency functions which are to be performed by some 28 Executive Departments and Agencies whenever the President of the United States declares a national emergency (as in defiance of an impeachment edict, for example). Under the terms of E. O. 11490, the President can declare that a national emergency exists and the Executive Branch can:

• Take over all communications media

• Seize all sources of power Take charge of all food resources

• Control all highways and seaports Seize all railroads, inland waterways, airports, storage facilities Commandeer all civilians to work under federal supervision

• Control all activities relating to health, education, and welfare Shift any segment of the population from one locality to another Take over farms, ranches, timberized properties

• Regulate the amount of your own money you may withdraw from your bank, or savings and loan institution.

All of these and many more items are listed in 32 pages incorporating nearly 200,000 words, providing and absolute bureaucratic dictatorship whenever the President gives the word.

Executive Order 11647 provides the regional and local mechanisms and manpower for carrying out the provisions of E. O. 11490. Signed by Richard Nixon on Feb. 10, 1972, this Order sets up Ten Federal Regional Councils to govern Ten Federal Regions made up of the fifty still existing States of the Union. Check out this book for the inside scoop on the \"secret\" Constitution.

SUBJECT: - \"The Proposed Constitutional Model\" Pages 595-621

Book Title – The Emerging Constitution

Author – Rexford G. Tugwell

Publisher – Harper's Magazine Press, Harper and Row

Dewey Decimal – 342.73 T915E

ISBN – 0-06-128225-10

Note Chapter 14

The 10 Federal Regions

REGION I: Connecticut, Massachusetts, New Hampshire, Rhode Island, Vermont. Regional Capitol: Boston

4

REGION II: New York, New Jersey, Puerto Rico, Virgin Island.

Regional Capitol: New York City

REGION III: Delaware, Maryland, Pennsylvania, Virginia, West Virginia, District of Columbia.

Regional Capitol: Philadelphia

REGION IV: Alabama, Florida, Georgia, Kentucky, Mississippi, North Carolina, Tennessee.

Regional Capitol: Atlanta

REGION V: Illinois, Indiana, Michigan, Minnesota, Ohio, Wisconsin.

Regional Capitol: Chicago

REGION VI: Arkansas, Louisiana, New Mexico, Oklahoma, Texas.

Regional Capitol: Dallas-Fort Worth

REGION VII: Iowa, Kansas, Missouri, Nebraska.

Regional Capitol: Kansas City

REGION VIII: Colorado, Montana, North Dakota, South Dakota, Utah, Wyoming. Regional Capitol: Denver

REGION IX: Arizona, California, Hawaii, Nevada.

Regional Capitol: San Francisco

REGION X: Alaska, Oregon, Washington, Idaho.

Regional Capitol: Seattle

Supplementing these Then Regions, each of the States is, or is to be, divided into sub regions, so that Federal Executive control is provided over every community.

Then, controlling the budgeting and the programming at every level is that politico-economic system known as PPBS.

The President need not wait for some emergency such as an impeachment ouster. He can declare a National Emergency at any time, and freeze everything, just as he has already frozen wages and prices. And the Congress, and the States, are powerless to prevent such an Executive Dictatorship, unless Congress moves to revoke these extraordinary powers before the Chief Executive moves to invoke them.

THESE EXECUTIVE ORDERS GROSSLY AND FLAGRANTLY VIOLATE THE INTENT AND PURPOSE OF ARTICLE 4 SECTION 3. THERE IS NO PROVISION IN THIS SECTION OR THE CONSTITUTION OF THE UNITED STATES FOR FORMING A REGIONAL STATE OUT OF A GROUP OF STATES! FURTHER, THESE EXECUTIVE ORDERS GROSSLY AND FLAGRANTLY VIOLATE THE 9TH AND 10TH AMENDMENTS TO THE CONSTITUTION!

By Proclaiming and Putting Into Effect Executive Order No. 11490, the President would put the United

States under **TOTAL MARTIAL LAW AND MILITARY DICTATORSHIP!** The Guns Of The American People Would Be Forcibly Taken!

The Hiv/Aids Fraud and Conspiracy

Reference in "Behold the Pale Horse" by William Cooper

In 1969 (3 years before the World Health Organization's request) the United States Defense Department requested and got $10 million to make the AIDS virus in lab(s) as a political/ethnic weapon to be used mainly against Blacks. The Feasibility program & lab(s) were to have been compleated by 1974 - 1975, the virus between 1974 - 1979. The World Health Organization started to inject AIDS-laced smallpox vaccine (Vaccina) into over 100 million Africans (population reduction) in 1977. And over 2000 young white male homosexuals (Trojan horse) in 1978 with the hepatitis B vaccine through the Centers for Diease Control/New York Blood Center. And now the AIDS virus is on the streets IN THE DRUGS

P L E A S E, W A K E U P!!

129 — Tuesday, July 1, 1969
SYNTHETIC BIOLOGICAL AGENTS

There are two things about the biological agent field I would like to mention. One is the possibility of technological surprise. Molecular biology is a field that is advancing very rapidly and eminent biologists believe that within a period of 5 to 10 years it would be possible to produce a synthetic biological agent, an agent that does not naturally exist and for which no natural immunity could have been acquired.

Mr. SIKES. Are we doing any work in that field?

Dr. MACARTHUR. We are not.

Mr. SIKES. Why not? Lack of money or lack of interest?

Dr. MACARTHUR. Certainly not lack of interest.

Mr. SIKES. Would you provide for our records information on what would be required, what the advantages of such a program would be, the time and the cost involved?

Dr. MACARTHUR. We will be very happy to.

(The information follows:)

The dramatic progress being made in the field of molecular biology led us to investigate the relevance of this field of science to biological warfare. A small group of experts considered this matter and provided the following observations:

1. All biological agents up to the present time are representatives of naturally occurring disease. and are thus known by scientists throughout the world. They are easily available to qualified scientists for research. either for offensive or defensive purposes.

* 2. Within the next 5 to 10 years. it would probably be possible to make a new infective microorganism which could differ in certain important aspects from any known disease-causing organisms. Most important of these is that it might be refractory to the immunological and therapeutic processes upon which we depend to maintain our relative freedom from infectious disease.

* 3. A research program to explore the feasibility of this could be completed in approximately 5 years at a total cost of $10 million.

4. It would be very difficult to establish such a program. Molecular biology is a relatively new science. There are not many highly competent scientists in the field. almost all are in university laboratories. and they are generally adequately supported from sources other than DOD. However. it was considered possible to initiate an adequate program through the National Academy of Sciences-National Research Council (NAS-NRC).

The matter was discussed with the NAS-NRC. and tentative plans were made to initiate the program. However. decreasing funds in CB. growing criticism of the CB program. and our reluctance to involve the NAS NRC in such a controversial endeavor have led us to postpone it for the past 2 years.

* It is a highly controversial issue and there are many who believe such research should not be undertaken lest it lead to yet another method of massive killing of large populations. On the other hand, without the sure scientific knowledge that such a weapon is possible. and an understanding of the ways it could be done. there is little that can be done to devise defensive measures. Should an enemy develop it there is little doubt that this is an important area of potential military technological inferiority in which there is no adequate research program.

It is reported articles that Robert Gallo, was the scientist who took the lead role in created this HIV/Aids Virus fraud for the global depopulation agenda and profit. It is the drug "Azt."

AZT, in full **azidothymidine**, also called **zidovudine**, drug used to delay development of AIDS(acquired immunodeficiency syndrome) in patients infected with HIV (human immunodeficiency virus). AZT belongs to a group of drugs known as nucleoside reverse transcriptase inhibitors (NRTIs). In 1987 AZT became the first of these drugs to be approved by the **U.S. Food and Drug Administration** for the purpose of prolonging the lives of AIDS patients.

It was reported by some doctors and scientists,stated that the T-cells(which is the human immune system) out number the Hiv virus from a **"100 to 1,"**when a person tested Hiv postive.
So if this claim of truth is fact, what really causes Aids. It is **(AZT)Azidothymidine,** and other hard narcotics drugs such as **crack cocaine,heroin,crystal methamphetamine**, and other hard illegal drugs.
These drugs suppress the immune system, and making the human body so acidity that the **(T-cells)** start to go against one another causing confusion in the body, which makes a person vulnerable to any disease.
In other reports, it states that people was intently getting misdiagnose as Hiv positive, so the doctors could persuade them to take the AZT medication, and collect profit and watch the person die in the process.

According to the "UNAIDS"and other sources

Using detailed statistical analysis,They found the most populous African countries of South Africa, Nigeria, Tanzania, Ethiopia, the Democratic Republic of the Congo, and Egypt would earn $2.2 billion in revenues for HIV/AIDS treatments for the years 2017 through 2021. In addition,they estimated the Middle East and North Africa and Sub-Saharan Africa regions would earn $4.3 billion, and worldwide revenues could be as high as $6.1 billion for that period.

Short List of Outputs

Outputs -- create controlled situations -- manipulation of the economy, hence society -- control by control of compensation and income.

Sequence:

1. allocates opportunities
2. destroys opportunities
3. controls the economic environment
4. controls the availability of raw materials
5. controls capital.
6. controls bank rates
7. controls the inflation of the currency
8. controls the possession of property
9. controls industrial capacity
10. controls manufacturing
11. controls the availability of goods (commodities).
12. controls the prices of commodities.
13. controls services, the labor force, etc.
14. controls payments to government officials.
15. controls the legal functions.
16. controls the personal data files -- uncorrectable by the party slandered.
17. controls advertising.
18. controls media contact.
19. controls material available for T.V. viewing
20. disengages attention from real issues.
21. engages emotions.
22. creates disorder, chaos, and insanity.
23. controls design of more probing tax forms.
24. controls surveillance.
25. controls the storage of information.
26. develops psychological analyses and profiles of individuals.
27. controls legal functions
28. controls sociological factors.
29. controls health options.
30. preys on weakness.
31.cripples strengths.
32.leaches wealth and substance.

Table of Strategies

Do This To Get This

Keep the public ignorant Less public organization

Maintain access to control point for
feedback

Required reaction to outputs (prices, sales)

Create preoccupation Lower defense

Attack the family unit Control of the education of the young

Give less cash and more credit and doles More self-indulgence and more data

Attack the privacy of the church Destroy faith in this sort of government

Social conformity Computer programming simplicity

Minimize the tax protest Maximum economic data, minimum
enforcement problems

Stabilize the consent Simplicity coefficients

Tighten control of variables Simpler computer input data - greater
predictability

Establish boundary conditions Problem simplicity/solutions of differential
and difference equations

Proper timing Less data shift and blurring

Maximize control Minimum resistance to control

Collapse of currency Destroy the faith of the American people in
each other

Diversion, the Primary Strategy

Experience has prevent that the simplest method of securing a silent weapon and gaining control of the public is to keep the public undisciplined and ignorant of the basic system principles on the one hand, while keeping them confused, disorganized, and distracted with matters of no real importance on the other hand.

This is achieved by:

* disengaging their minds; sabotaging their mental activities; providing a low-quality program of public education in mathematics, logic, systems design and economics; and discouraging technical creativity.

* engaging their emotions, increasing their self-indulgence and their indulgence in emotional and physical activities, by:

o unrelenting emotional affrontations and attacks (mental and emotional rape) by way of constant barrage of sex, violence, and wars in the media - especially the T.V. and the newspapers.

o giving them what they desire - in excess - "junk food for thought" - and depriving them of what they really need.

* rewriting history and law and subjecting the public to the deviant creation, thus being able to shift their thinking from personal needs to highly fabricated outside priorities. These preclude their interest in and discovery of the silent weapons of social automation technology.

HAARP MACHINE
"High Frequency Active Auroral"
Reference **"Angels Don't play this Haarp"**

Environmental Warfare?

The U. S. Government has a new ground-based "Star Wars" weapon which is being tested in the remote bush country of Alaska. This new system manipulates the environment in a way which can: • Disrupt human mental processes. • Jam all global communications systems. • Change weather patterns over large areas. • Interfere with wildlife migration patterns. • Negatively affect your health. • Unnaturally impact the Earth's upper atmosphere. The U.S. military calls its zapper HAARP (High-frequency Active Auroral Research Project). But this sky buster is not about the Northern Lights. This device will turn on lights never intended to be artificially manipulated. Their first target is the electrojet - a river of electricity that flows thousands of miles through the sky and down into the polar icecap. The electrojet will become a vibrating artificial antenna for sending electromagnetic radiation raining down on the earth. The U.S. military can then "X-ray" the earth and talk to submarines. But there's much more they can do with HAARP. This book reveals surprises from secret meetings. PROJECT CENSORED - a prestigious panel of journalists - judged HAARP to be in the top ten under-reported news stories of 1994. POPULAR SCIENCE - As a front-cover story, HAARP began to be revealed in September, 1995. This book is - the rest of the story. The High-frequency Active Auroral Research Project is The HAARP that Angels Don't Play.

The U.S. Government has been using this machine to make hurricanes,tornadoes,forest fire and earthquakes. So we as a people, don't be surprise when a natural disaster suddenly occurs and destroys homes and cities. It is a global depopulation agenda. And if you think this is false, here is a statue in Georgia, saying "Maintain Humanity under **"500,000,000"**

For an even more detailed and revealing 45-minute History Channel documentary on HAARP and other secret weapons used for electromagnetic warfare,Here are two quotes from the History Channel documentary:

"Electromagnetic weapons … pack an invisible wallop hundreds of times more powerful than the electrical current in a lightning bolt. One can blast enemy missiles out of the sky, another could be used to blind soldiers on the battlefield, still another to control an unruly crowd by burning the surface of their skin. If detonated over a large city, an electromagnetic weapon could destroy all electronics in seconds. They all use directed energy to create a powerful electromagnetic pulse."

"Directed energy is such a powerful technology it could be used to heat the ionosphere to turn weather into a weapon of war. Imagine using a flood to destroy a city or tornadoes to decimate an approaching army in the desert. The military has spent a huge amount of time on weather modification as a concept for battle environments. If an electromagnetic pulse went off over a city, basically all the electronic things in your home would wink and go out, and they would be permanently destroyed."

Here are some "Haarp"locations around the world.

ChemTrails

It is a biological agent of toxins release in the air by jet planes, which is use to pollute the air,water, and the land. It is these toxins that destroy crops ,and disrupt the human brain, with unbalance thinking, for mind control over the masses. These corrupt governments around the world used this as another biological weapon against the people. The chemicals and toxins such as heavy metals mixed with other harmful substances alter the the human brain waves, causing people to become very irritated and angry. Now the manipulation of mind control begins.

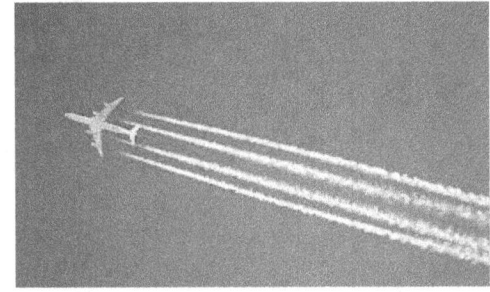

Chapter 9

Ancient Kemet and The Mdw Ntrch

Ancient Kement civilization dates back to over "**4,700 yrs**" B.C.E "before common era"

The word **(Kemet)**, which we know as **(Egypt)** today in modern era. It was originally spell **(KMT)** meaning **"Black Community"** not *Black Land*. In ancient Kemet, The language they spoke was called the **"Mdw Ntchr"** which meant **"Mother Nature"** The ancient people of the **"Hapi Valley"** which we know today as the **"Nile Valley"** did not use any vowels in their language until the last Golden Ages of their Rule. So therefore, We will see the words spell out with just consonants and no vowels, as in **(KMT)** **"Kemet"**, also used another name which was **"TMR"**(Ta Merri) Meaning the **"Beloved Land"**

This is the Ancient Kemet Text on the pyramid wall name "KMT" or "Kemet"

This is the Kemetic Text "TMR" (Ta Merri) The "Beloved Land"

The word pyramid was **(Mrk) "Merku"** and was built with advance sacred geometry, In the sense of duplicating the heavens, such as the sun, moon and stars. **"As Above, and So Below"**

Here is The Kemetic Alphabet

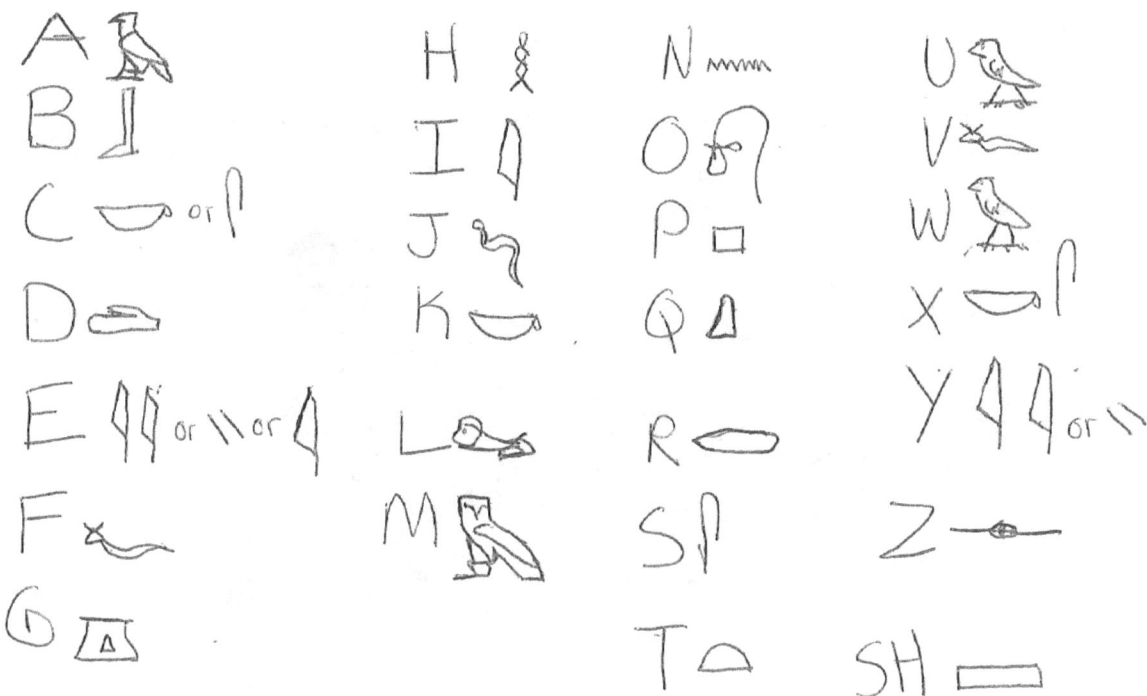

*Not all letters represent single letters, some represent words

1. Ideograms- were used to write words they represented
2. phonograms- were used to spell out the sound out, the words they represented usually had no relation to the word

Here is a timeline of the dynasties in Ancient Kemet
4,700 years B.C.E, Before their was a Jew/Hebrew, Greek and Roman.
So by this being proof on the pyramid texts,temples, and statutes in the
(Hapi)Nile Valley. So How is that these people, The Jews,Hebrew,Greeks
and Romans become the origin of this culture of knowledge. This is my
thing,can anyone in their Torah, Holy Bible, and Holy Q'uran show any proof
that the people in them books ever existed. We see clear evidence in Ancient
Kemet(Egypt) that these people existed in those times.

Ask your self this question
"Do you want facts or lies"

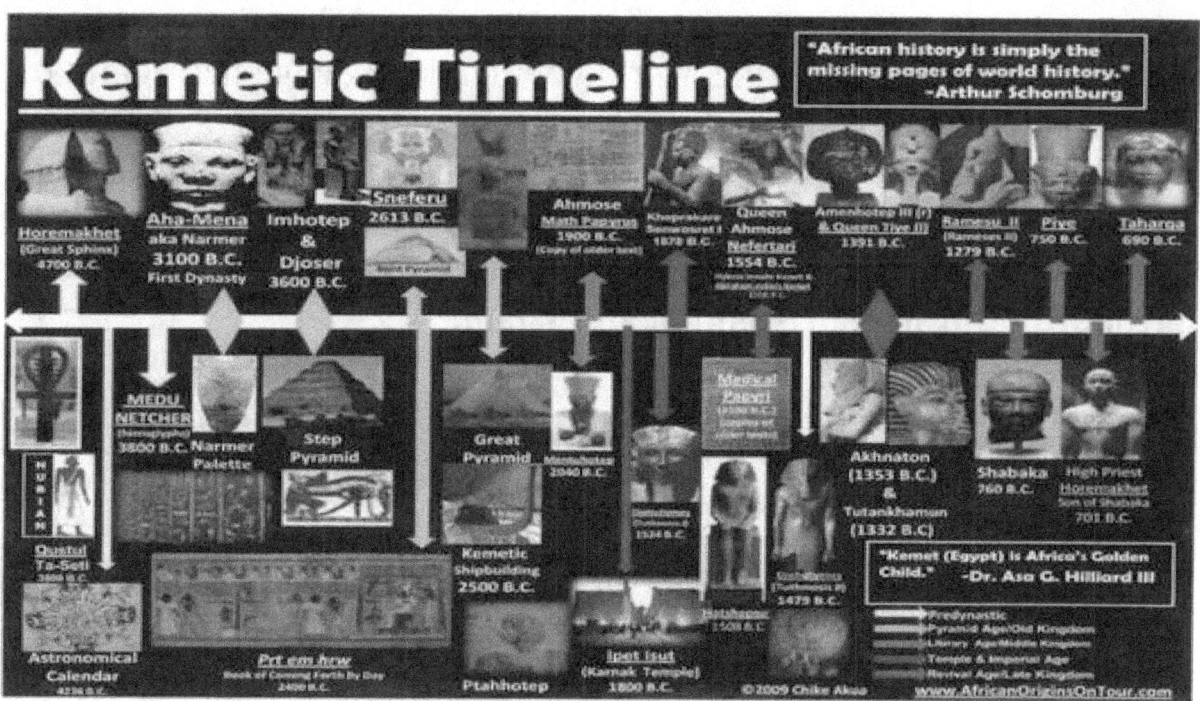

The people in ancient Kemet foundation of culture science ,came out of the
land of **Kush or (Kash) "The Old Kingdom"Today known as
"Ethiopia"**The science of what they knew already was into form. It was
pyramids already built hundreds of years before they introduce the world to
civilization.

Now, was is the difference between
"Culture" and *"Civilization"*

Culture is a way of life, such as understanding the time to harvest for food for the sake of survival, Having a sense to know your purpose here on earth, connecting with nature, an expression that relates to your everyday life, and understanding moral values of family and nature that is practice on a consisting basics.

Civilization is when you elevate that culture and have organization of science, writings, languages, maps, math system, universities of studies of different fields,laws and a government. So therefore, Civilization is the advance form of culture organize to its greatest heights.

If you want to believe it or not. The Fact is that over **1/3** of the cultures, religions and civilizations in this world got their knowledge that comes out of ancient Kemet. And if you think this is a lie, the ancient text on the pyramids can prove this with the date of the time period shown on the Kemetic timeline before the **Sumerians ,Phoenicians, Mobaties (Jews) or (Hebrews),** same scholarship of Kemetic science, **Greeks, Romans, Judaism, Buddhism, Christianity, and Islam.**

I suggest the people research the work of (Dr. Yousef Ben and Mfundishi)

Now, I will give you some Food for Thought.
Sound, was one of the ways that ancient Kemet built the pyramids, with a formula using acoustic **sound.** Inside the pyramids it was sections rounded off in a circle of **360 degrees**, and when they hit a certain pitch, they could make an object weightless and break the laws of gravity with will.

C = 2 π r or radius, which is 3.14

Stolen Legacy

This is the original Trinity from **"Auset, Ausar, & Heru"** is were most religions get this knowledge from in there doctrine or belief system,when pertaining to the Trinity.

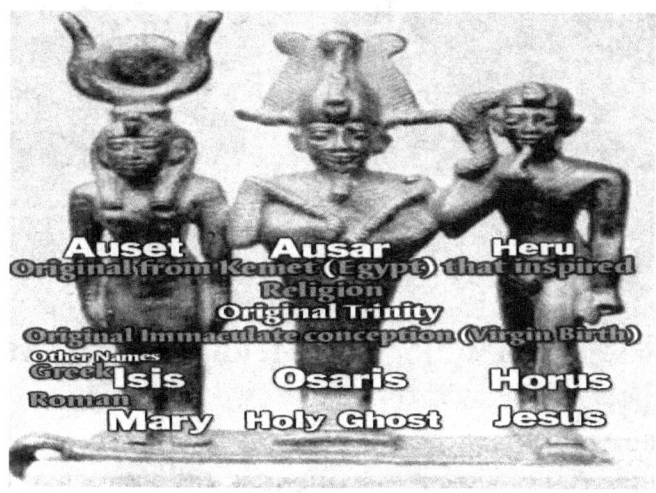

Now here is "Auset and Heru" the two Deity's of the origin of the immaculate conception"The Virgin Birth"

Here is Ma'at depicted as a woman with ostrich feathers

She was a deity in ancient Kemet who inspire **"Tehuti" or "Djhwty"** on a spiritual sense "the feminine aspect of life, and representing **"Truth,Justice,Righteousness,Reciprocity,Order,Harmony and Balance."** bringing fore the laws of nature into a constructed sequence in the physical form or underworld. Without the inspiration of "Ma'at" the laws today of any kind of government wouldn't not even play a part in modern day time.

Now here is "Tehuti" or "Djhwty" depicted as a Ibis bird recording and writing the scribes of the inspiration of Ma'at from the feminine aspect of nature.

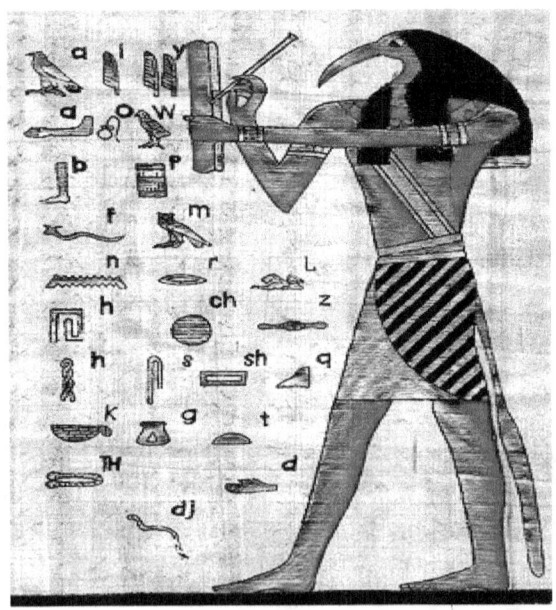

He was a master of writings,**(The scribes)**, the science of art in many forms of expression and recorded the experiences of nature, and brought fore the laws of the land over **2000 years** before the fairy tale story of **Moses** in the **Hebrew Bible orTorah.**

Here are the (42 Laws) of Ma'at or the (42 Konfessions) of Ma'at)

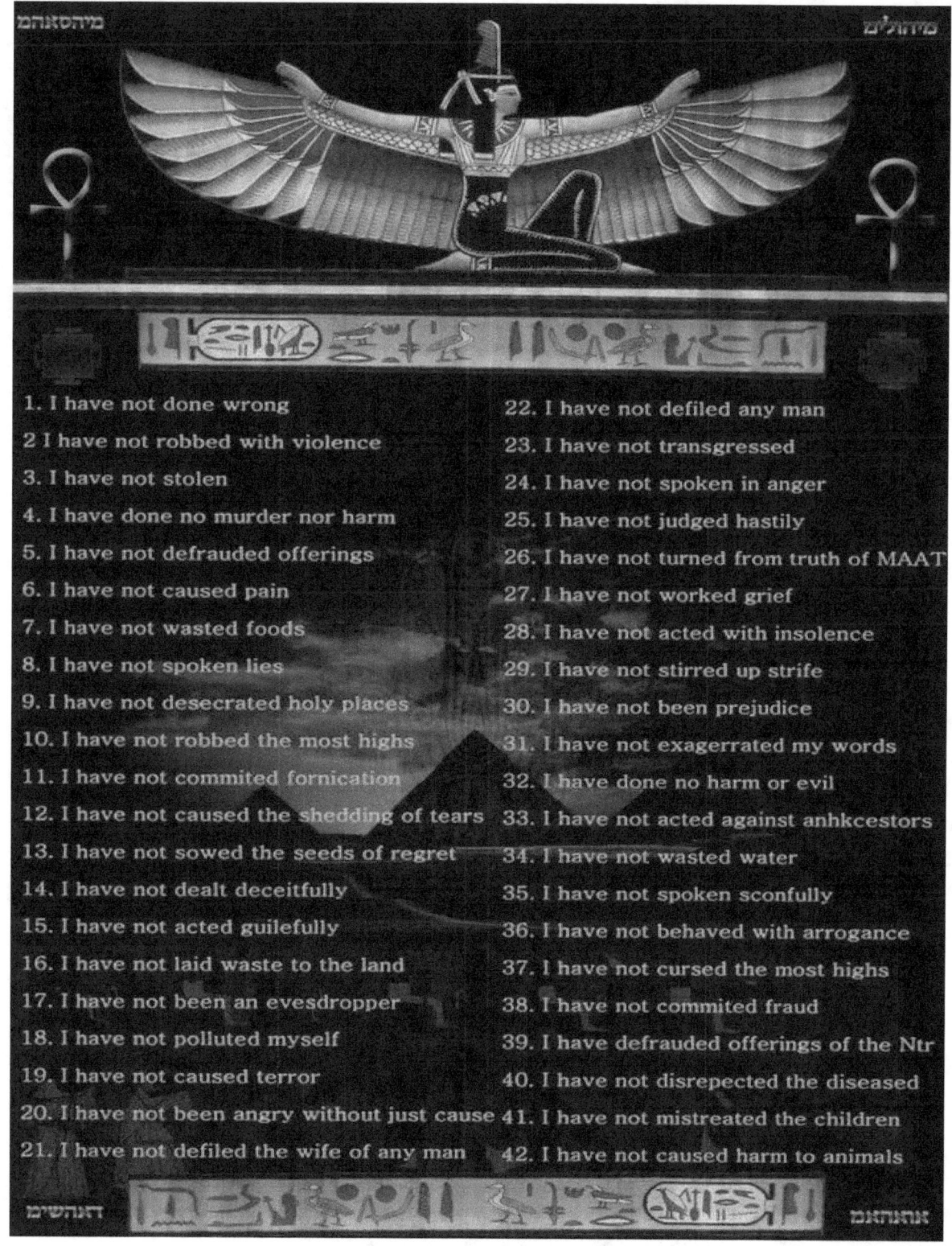

1. I have not done wrong
2 I have not robbed with violence
3. I have not stolen
4. I have done no murder nor harm
5. I have not defrauded offerings
6. I have not caused pain
7. I have not wasted foods
8. I have not spoken lies
9. I have not desecrated holy places
10. I have not robbed the most highs
11. I have not commited fornication
12. I have not caused the shedding of tears
13. I have not sowed the seeds of regret
14. I have not dealt deceitfully
15. I have not acted guilefully
16. I have not laid waste to the land
17. I have not been an evesdropper
18. I have not polluted myself
19. I have not caused terror
20. I have not been angry without just cause
21. I have not defiled the wife of any man

22. I have not defiled any man
23. I have not transgressed
24. I have not spoken in anger
25. I have not judged hastily
26. I have not turned from truth of MAAT
27. I have not worked grief
28. I have not acted with insolence
29. I have not stirred up strife
30. I have not been prejudice
31. I have not exagerrated my words
32. I have done no harm or evil
33. I have not acted against anhkcestors
34. I have not wasted water
35. I have not spoken sconfully
36. I have not behaved with arrogance
37. I have not cursed the most highs
38. I have not commited fraud
39. I have defrauded offerings of the Ntr
40. I have not disrepected the diseased
41. I have not mistreated the children
42. I have not caused harm to animals

Here is the Ankh "The symbol of life"

The Ankh symbol is represented as the female vagina in its shape and form. Giving the glory to the power of the womb of **"wombman "or "woman"** It also reveal the creative force, from the "feminine and masculine" aspect of both woman and man coming together as 1 to create another life

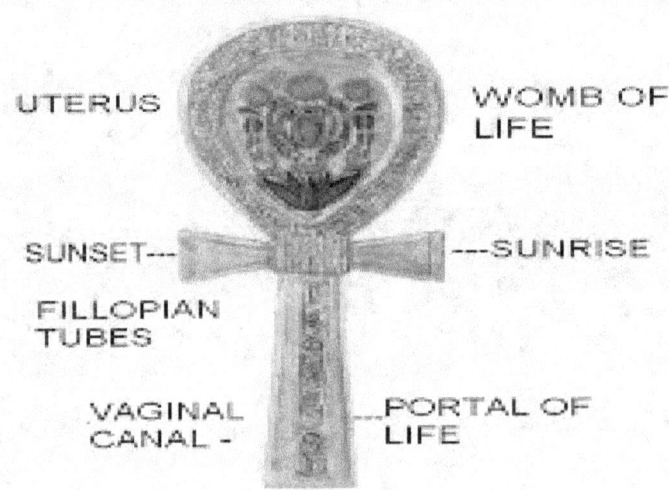

It is a symbol that was used all threw Ancient Kemet, But its origin dates back thousands of years before Kemet came into play. The people of ancient **Kush or Kash** already had establish this science of the Ankh symbol, in terms of understanding the **wombman** is were the creation of life begins.

Now, We in modern times we refer the **"Serpent or Snake"** as something bad.
But that is a false expression. When we take the to time and study Ancient Kemet,
Everything they did was a connection to nature itself. That is why the language they spoke was called the **"Mdw Ntchr"(Mother Nature)**
That is why you will see images of them depicted as a animal comparing their attributes of that animal strength in nature.

So we need to understand **"Systematic Thought and Symbolic Thought"**

What is Systematic Thought?
It is the original idea of that thought ,the direct point and focus. Pertaining to, fixed or plan system, combine in whole

What is Symbolic Thought?
It is an expression of comparison or similarities to the original thought, describes with metaphors of incorporating what that original thought literally meant

The Deity's of Ancient Kemet wore the **"Seperent Crown"** or **"Uraeus"** on top of the forehead which symbolize the **"Pineal Gland"**. The pineal gland is the **"7 chakra"** or **"Crown Chakra"** representing the activation of the high stages of consciousness. **Now what is the meaning behind that**. Because the **snake or seperent,** it spirals when it move, just like the (**Kundalini**) energy move in the body to activate the higher brain senses. **It represents High wisdom, High Honor and Divine Power**

The **Systematic Thought** is having **"High Wisdom or Divine Power"**

The **Symbolic Thought** is that the way the serpent move in a spiral just like the Kundalini Energy to reach high stages of consciousness to have high wisdom and that divine power of light.

THE "URAEUS" OR "SERPENT CROWN"
SHOWN ON EGYPTIAN AND NUBIAN ROYALTY
ALSO KNOWN AS THE NAGAR OR THE NEGUS.

AMONGST MANY INDIGENOUS AFRICAN CULTURES
THE URAEUS IS A SYMBOL OF WISDOM, POWER
AND PROTECTION.

Chapter 10
Esoteric Knowledge,Symbols and Occult Science

Esoteric is a small group of people only intended to understand.
Occult is Hidden Culture.

We need to know that these corrupt governments, and secret societies such as Freemasons, Skull & Bones, Ordo Tempi Orientis, and the Bilberg Group etc. They have hijack ancient science and knowledge and preserve the knowledge for themselves to control the Masses.
Their agenda is clear, control, and keep the public ignorant, as slaves, **"The cattle".**
This is one of the most popular symbols, which is the Moorish seal. This symbol has been distorted by secret societies. Many believe it is the illuminati. Which is fact, when we is dealing with the truth science of it. The word "illuminati", derived from the word "illuminatus" and "illuminare" in Latin. Which means to be "enlightened" or to have light. So when you activate the higher brain senses, and rise to high stages of consciousness, that is when you become enlightened and you is now illuminati.

The eye represent, The pineal gland,Which is the eye of Heru. The Kemetic Deity**(Heru)** symbolize the crown chakra, of high stages of consciousnesses,The 3 points, represent the Trinity of self, Knowledge,Wisdom and Understanding. It is a symbol of civilization.
The secret societies separated the top of the pyramid eye, detailing that they will suppress the people from achieving high stages of consciousness. They distorted the Moorish seal, confusing the masses that the culture science of the moors belong to them.

Now when we examining this symbol below, we see the star spangle banner and the red ,white and blue stripes with the 50 stars. What is the significance of that. The 50 stars represent the orbiting period of Canis Minor of the star Sirius B, which is 50.1 and is consider to be the dog star, which was praised by the **"Dogon"** The Ancient Kemetic tribe who study the constellation, which is a **"Nebula"** a ball of gas that birth stars, and said that the **"Nebula"**is were our sun was birth from. So it is no surprise why its nickname is the dog star, in Canis Minor. The star Sirus B. colors are red,blue and white, so the U.S. Adopted that to be their colors of their Star Spangle banner in paying high regards to the star Sirius B.

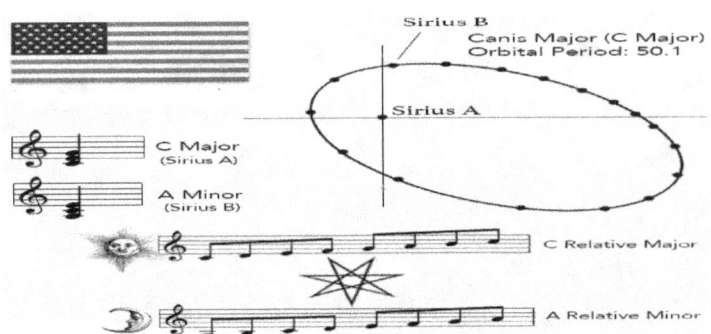

Now, lets examining this image and what does it symbolizes, many people believes it is the devil,but that is distorted by the secret societies to scare people away from the true science, so the people become so afraid and want used it. So the occult groups like the Freemasons used that tactic to control the masses with fear and distorting the images.

"The name of the Templar **Baphomet**, which should be spelt kabalistically backwards, is composed of three abbreviations: **Tem. ohp. AB., Templi omnium hominum** pacts **abbas**, "the father of the temple of peace of all men".

This image is consider to be a **Hermaphrodite** meaning male and female.

The **horns** on its head represent fertility, as it is shape in the form of the crescent moon representing the womb-man, who is fertile to bring fore creation of life.

The **2 snakes** represent the rising of the kundalini energy, from the root chakra to the crown chakra.

The **5 point star** represent the connection of the **5 senses** of the human body.

In latin, the word **"Solve and Coagula"** means

Solve et Coagula s the essential alchemical process. "Solve" or **"solutio"** refers to the breaking down of elements and **"Coagula"** refers to their coming together. In the process of transmuting or transforming energy. This contained both literal and metaphorical meaning.

*****Transforming energy from your (lower self to the higher self) "Feminine and Masculine" Left and Right**

The **"Two-finger salute"** Two fingers on the right hand point up and two on the left hand point down, meaning **"as above, so below"**.

In **Greek** , the word Devil comes from the word **(Deus,Dios, or Diety)**
Meaning Divine or the thought of god, as the giver of water
Dios,DI(donate) OS(water as in the ocean.

So when we take the time to reveal the science of this image and symbol, It is nothing scary about it, It is simply revealing the science of life, right in front of your eyes for you to used.

Who is **Satan**. The word Satan comes from the word **"Set"(Darkness)**in ancient Kemet, The Kemetic deity who oppose the Kemetic deity **"Heru"(Light)** denoting the science of polarity. The Root word of Satan is **"Sat"** which originates from the word **"Set"** meaning darkness,desert,chaos,storms and representing the planet Saturn.

The ideal is the science of polarity, The opposite between the two. So when it was day **"Heru"**the sun deity brought light, and when it was night, **"Set"** had conquer the battle in darkness. The reason is that **"Saturn"**, is the most visible planet furthest from earth that the human eye can see. It it was used as a bright star, consider to be the sun at night, and used for processional time keeping for agriculture.

Now, lets look at the number **13.** Many people says it is a unlucky number. What is the significance of this number people is afraid of. The occult groups has suppress this number, which is suppressing the woman or womb-man. The number **13 represent the 13 cycles of the moon** and the woman denotes the moon, in representing **fertility**. The reason is the crescent moon is symbolic to the any animal with horns, So in ancient times, they used that as a symbol of **fertility.** So the occult groups suppress these things to keep the woman off balance and their natural cosmic rhythm. The woman suppose to be in tune with the rhythm of the moon with **13 cycles or periods, menstruation a year, Not 12.**
13 cycles x 28 average days of the moon, gives you **364 days** out the year.

The secret societies agenda is to keep the woman off rhythm, because they know if the womb-man rise up into her cosmic natural rhythm with balance, the womb-man will take back the earth and lead the way to restore **"Ma'at"(Truth,Order,Justice,Harmony and Balance".**

What we see below is how the occult express the number 13, but will never reveal the truth science and meaning behind these symbols, So it is up to you to do your research,study and just used your common sense.

Using the language of Gematria

"mcdonalds" in the *English Ordinal* system equals 85 (13+3+4+15+14+1+12+4+19), which reduces to **13**, which reduces to **4**
Expressing the Lunar Calendar, **13 cycles of the moon x 4 weeks of the month, =52 weeks, which = 364 days**
13 x 28=364 days

Also the U.S used 13 stars, for the 13 colonies, denoting the lunar Calendar

So what is so bad about the number 13, It is just a number denoting the lunar calendar and how it connects to the woman on a cosmic level of nature.

What is the reason for **religion**. The word **religion** comes from the word **"ligare"** in Latin meaning to bind, tie tight and tie down. In other words, the meaning is saying to limit something or control.

Religion was presented to the people for control and to have the masses believing in some type of mystery god, that is going to come down from the sky and save them from all their sins. When we go deeper, We will find out it is **"Dogma"** which is a doctrine of a belief system, such as a pastor,preacher, and high priest etc. These people of **"Dogma"** in these religious institutions know the science and hidden truth. The idea is to sell the people the lies and keep people far from the truth for control. So they can continue to mentally enslave those who is trap on some type of belief system and continue to rob the people for their money, time and energy. Have you ever wonder why, you in the struggle to pay your bills and the preacher man is riding around in a new Mercedes Benz, But he will tell you to continue to pray for the mystery god, while you is poor and still struggling hoping that he will come to changes things. These people who follow belief system is steady wasting their time and money waiting for that spook in the sky that is never going come. When studying religion,We must understand systematic thought and not get trap on the symbolic thought. In these holy books, I find out that it is base all on astronomy, astrology, and cosmology. These belief systems neglect science and which is the foundation of life. This is very critical for survival. So the people in **"Dogma"**condition the people to sway away from the truth science of life and pray to false prophets and the spook in the sky. When I study these biblical books, I find out that these holy books are plagiarize stories and 1/3rd of all their knowledge comes out of ancient Kemet and the Sumerian text. The ideal is to go to the source were the information or knowledge comes from and understand systematic thought first, before you understand symbolic thought. **"Research Dr. yosef Ben.**

What is the **"Mark of the Beast"**. In the King James version of the Holy Bible, its speaks

(666)

Revelation 13:18- Here is wisdom. Let him that hath understanding count the number of the beast: for it is the number of a man; and his number [is] Six hundred threescore [and] six.

What is the meaning of this scripture. We need to understand that the holy bible is built on astronomy,astrology, and cosmology.

All it represents is the **(6 electrons, 6 protons, and 6 neutrons)** in the atom of the element **"Carbon"**, **"666"**

which is necessary for the creation of man. **(As shown here below.)**

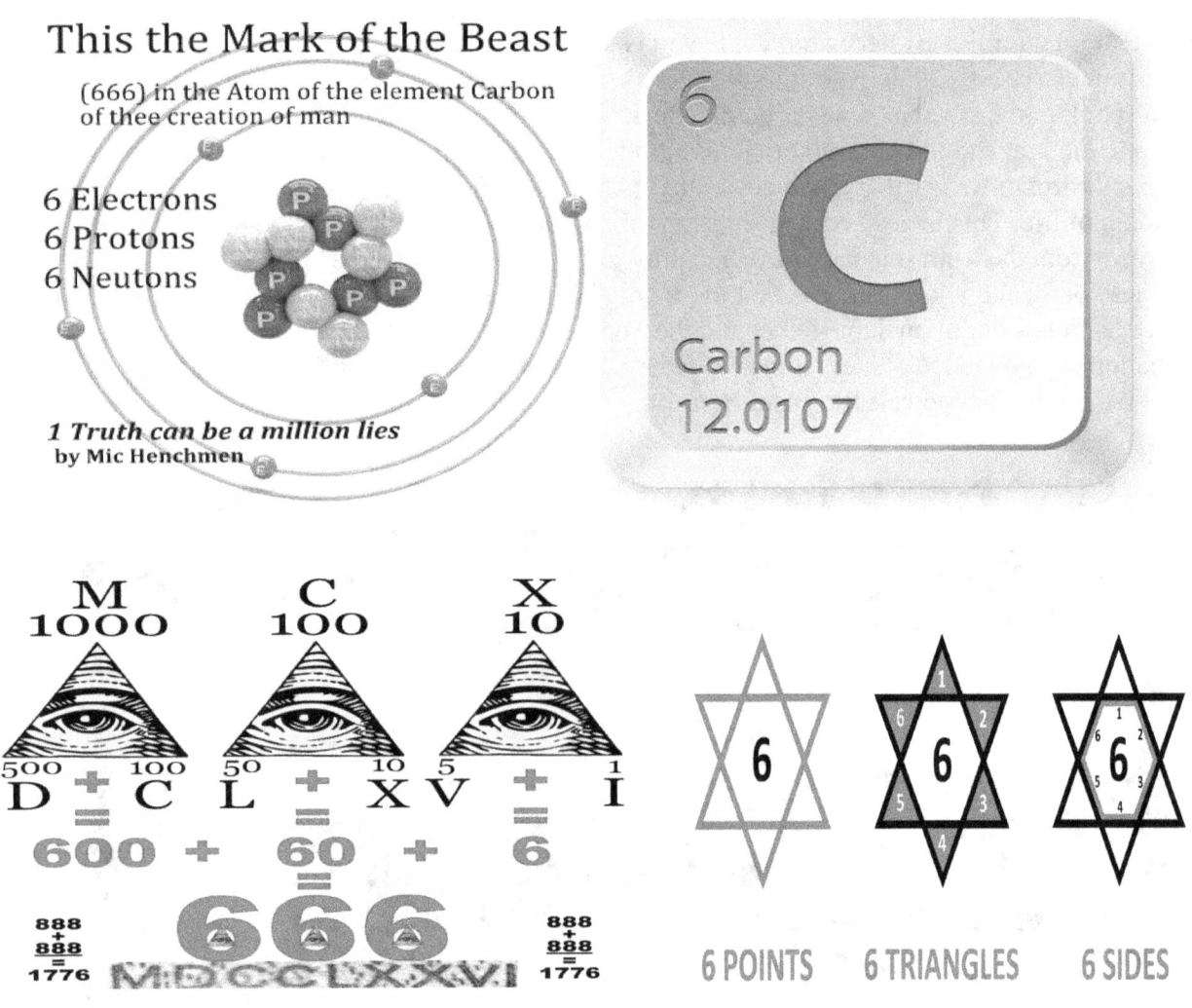

This the Mark of the Beast

(666) in the Atom of the element Carbon of thee creation of man

6 Electrons
6 Protons
6 Neutons

1 Truth can be a million lies
by Mic Henchmen

6
C
Carbon
12.0107

M
1000

C
100

X
10

D ± C   L ± X   V ± I

600 + 60 + 6

666

888
+
888
=
1776

MDCCLXXVI

888
+
888
=
1776

6 POINTS   6 TRIANGLES   6 SIDES

As Above, So Below
Trinity(3)

Chapter 11
Interesting Things to Know

Did you know that 1 minute of anger can shutdown your immune system up to 6 hours.

Did you know that the Pope of Rome is the Boss of all these secret societies, the U.S. And other world governments. The Vatican hold the esoteric and occult science which was stolen by high rank priest, and has hijack world knowledge. They have it hidden in the vault in the Vatican. They preserve this science to control and manipulate the masses.

Did you know that King James was Prince James first, and full name was **"James Charles Stuart"** A so call black man, who higher over 50 Europeans scholars to write the bible in his version.

Did you know a man name Aleister Crowley is one of the main reason why it is so many homosexual activities in Hollywood and the world today. He has a book call the book of law, which teaches people to go out their nature,and to perform sick rituals such as having sexual intercourse with their mother,father,sister,brother and young children. He also speak about engaging into acts of having human feces and urine smeared on the partner of the act and they will rewarded with materialistic thing in life in the sick ritual practice. His ideal is to take people out of their divine nature and follow a confuse way of life.

Did you know that if the male right side of their body is more activated during sex, that most likely the female will produce a male child, and if their left side of the your body is more activated during sex the female will have a female child.**"Right"** is masculine, **"Left"** is Feminine. "Understand the principles of polarity.

Did you know that the Earth is furthest distance from the sun in the summer time, and the earth is at its hottest peak, on July 4th its distance is 152,100,000km, and on January 3rd its distance is 147,300,000km

Did you know that the first bibles was the **"Septuagint"**transliterated by the Greeks from Hebrew.

Did you know that dark hue people**(Carbonated people)** or the so call black people, absorbs the ultra violet rays from the sun and automatically transform it to vitamin d and a source of energy, But those who lack **"carbon"**, such as the pale or the so called white people,the ultra violet rays, becomes toxic to their body which causes skin cancer and many others health problems.

Did you know that Europe is not a continent and is the Northwest part of Asia. Every other continent Start with an A, such as North **(America)**,South **(America),Africa,Australia, and Asia.** Why isn't Europe?The Caucasians made a border line separation in of north west Asia, for the regions that they inhabit in Asia.
In the Europeans public text books in school, that were calling it Eurasia. That why they develop the Mercator projection map,in the early 1500s to change the geography of the land masses, to suppress that truth. In 1974 they develop the Peters projection map with more accurate size of the land mass.

Did you know that the first Pope of Rome was a so called black man, of dark hue of Moorish descent.

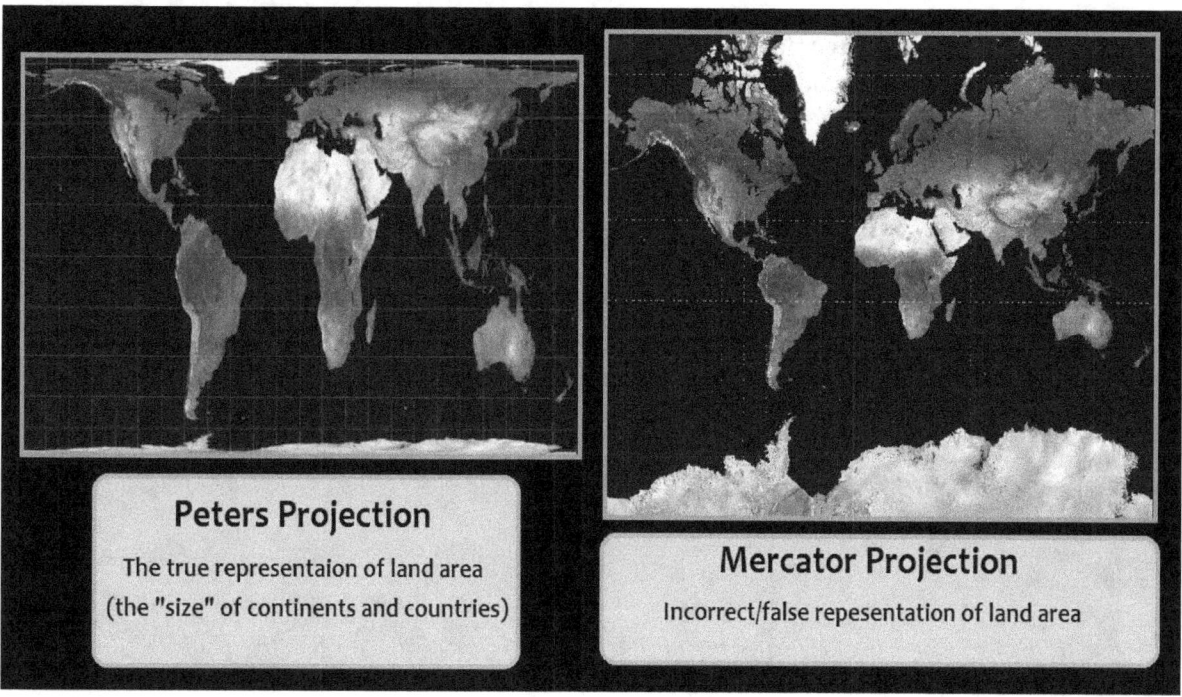

Peters Projection
The true representaion of land area
(the "size" of continents and countries)

Mercator Projection
Incorrect/false repesentation of land area

Did you know that the word **"Hebrew"** comes from the word **"Ibri "**in Hebrew, meaning the those who cross the river,and the word **"Israel"** meaning ascension. So when we speak of **Hebrew Israelite**, All it is saying those who cross the river and ascending to high stages of consciousness.

Did you know that the story of Santa Clause or Saint Nicholas was a Moorish man of dark hue. He was a bishop, who played a dominant role in the mid 1400s and major influence in Europe.
He died December 6, AD 343 in Myra and was buried in his cathedral church. So on the pagan holiday,of Christmas, people will say **"Merry"** Christmas. The word **"Moor"** comes from the word **"Mer"**Mer to (Mer)-ry.

THE ART OF THE DAY DON'T LIE! BUT, WHITE MEN HAVE LIED FOR CENTURIES
The Original Santa Claus
Was A Moor Born In Turkey

Did you know that the word **"Mermaid"** is pertaining to the European Caucasians women, who were servants for the Moors.**"Mer"** maid.

Did you know that to join and become initiated in the secret societies such as freemasonry, you must be blind fold, on your knees with a sword above your head swearing to secrecy not to tell the secret. Then you must perform a homosexual act to the high priest, a ritual of finding the **"short and curly"**.In other lodges, they will make the person ride a goat in a circle blindfold, imitating what the Moors did to the European Caucasians when they had them as prisoners transporting them to different tribes of the Moors. That is why in freemasonry, they have that saying is you **"Riding that Goat"**
So they prepare the people in college in these fraternities & sororities, with the brother and sister hood before they get to the real deal. **My people, you must wake up.**

FIG. 8.

Master.   Altar.   Candidate.   Conductor.
CANDIDATE TAKING THE OATH OF AN ENTERED APPRENTICE.

" Kneeling on my naked left knee, my right forming a square; my left hand supporting the Holy Bible, square, and compasses, my right resting thereon."

RIDING THE GOAT.

Did you know that the biggest secret in the Freemasonry and other secret societies is suppressing the truth history of the Moors, which is the so called black people. They trick the indigenous/native people in America to believe that they were slaves that came from Africa, so the social engineering have the so called black people thinking that they are not originally from America and that they are foreigners. So as long as the lie can continue and the true history is suppress, the Europeans can continue to have the nation of people thinking that they are the sovereign people of the land in America, which in reality they are the true foreigners.

Did you know that the European Freemasons will take a **"so called black man"(Moor)** bum off the streets and pay them some money,then blind fold them, then bring them in the lodge and perform a ritual with the master mason telling those under rank that the **"so called black man"(Moor)** is their father and they should always suppress their father,never let their father know who he is. **That is why the true history of the Moors is suppress and is the biggest secret in the European Freemasonry lodge.**

Chapter 12

The MichaEl Scribes of (36)

1.We the people, is the law which govern our self according to the law of nature, **"I am self law am Master"** (I Am, I Am)

2.Good karma, Bad karma is symbolic to the domino effect. What goes around comes around.

3.The law of physics states neither energy is created nor destroyed, So in that sense that is stating that **"Life is Eternal"**.

4.**DNA** is a center of stored information past on by your ancestors, So to be a **"Genius"**, you must tap into your **DNA,** and activate your consciousness.

5.If you wish to understand the universe, you must take the time to study yourself.

6.Ignorance is one of the worst diseases and can only be heal with spiritual truth.

7.If negative energy didn't exist, the world wouldn't have no balance.

8.If you feel something is necessary to be done, so do it.

9.Religion has always enslaved the minds of people, so therefore those who follow a doctrine based upon a belief system, will never be spiritually free.

10.Freedom in this society today will never be given to you, it must be taken.

11.Words have power because it has frequencies and vibrations, so if it's something positive, speak it into existence.

12.The key to be connected to nature is to be in tune with your inner self to have love.

13.Your foundation is knowledge, so you need to **know- the- ledge**, so you want fall off edge.

14. The more you study and learn, the more it is to know **(knowledge is infinite).**

15. When you neglect nature you is neglecting the God that is within you.

16. Time can only exist if matter is created, so therefore, matter is time and time is matter.

17. A ignorant person is quick to speak upon theories and lies, but is slow to listen to the facts and the truth.

18. Racism only exist in the world of social engineering, there is only on race and that is the human race.

19. One truth can beat a million lies any time of day.

20. The world is a cycle that will have many changes, so if you is not prepare for change, that change will eventually change you.

21. It states in the **"Seven Hermetic Principles"**, All is mind and the universe is mental, so if you got a mind, you need to learn how to use it.

22. This land here in the Americas was already populated with over 150 million people, so how can a European settler discover something when people was already there.

23. If you put water in a cup, it becomes the cup, if you put water in a bowl, it becomes the bowl, so the idea is to be free like water.

24. When it comes to being conscious, that person must understand to **"let go their ego"** because ego will enslave the mind, and suppress the spiritually growth of ones consciousness.

25. Like the earth, we must be grounded for the sake of survival.

26. **Everything is Energy,** and **Energy is Everything,** so therefore everything in the universe is connected to 1 source, which is **"Energy" (Everything is Everything)**

27. Like the mighty falcon, you must oversee this world with the vision of **"Heru"**.

28. When it comes to making a critical decision, you must understand

the vibrations you is feeling and listen to your inner self.

29. To believe, **"is to be left"**,so the only thing a person have when they believe is just hope.

30. You must know your purpose of life, or you will never understand the true reason why you are here.

31. Any decision that you may make in your present,will determine the effect of your future, so you want to make wise judgments, to get a wise result.

32. Once you become more spiritual, the physical world will become simple to you.

33. In Latin, the word **"man"** means **"to think"** so therefore, the meaning is detailing that man is to be on a conscious path within his or hers journey in life.

34. A wise man take the time to listen, while a foolish man speaks irrational and acts ignorant.

35. Nature does not discriminate,Nature can care less how much money you got and what color you are. Nature is going to be Nature, So we as a people must be able to do the same.

36. **Before it was light, it was darkness, before it was darkness, it was energy, before it was energy, it was always energy.**